CHINESE NATIONAL SECURITY DECISIONMAKING UNDER STRESS

Edited by
Andrew Scobell
Larry M. Wortzel

September 2005

The views expressed in this report are those of the authors and do not necessarily reflect the official policy or position of the Department of the Army, the Department of Defense, or the U.S. Government. This report is cleared for public release; distribution is unlimited.

*****

Comments pertaining to this report are invited and should be forwarded to: Director, Strategic Studies Institute, U.S. Army War College, 122 Forbes Ave, Carlisle, PA 17013-5244.

*****

All Strategic Studies Institute (SSI) monographs are available on the SSI Homepage for electronic dissemination. Hard copies of this report also may be ordered from our Homepage. SSI's Homepage address is: *http://www.carlisle.army. mil/ssi/*

*****

The Strategic Studies Institute publishes a monthly e-mail newsletter to update the national security community on the research of our analysts, recent and forthcoming publications, and upcoming conferences sponsored by the Institute. Each newsletter also provides a strategic commentary by one of our research analysts. If you are interested in receiving this newsletter, please subscribe on our homepage at *http://www.carlisle.army.mil/ssi/newsletter.cfm*.

ISBN 1-58487-206-3

CONTENTS

FOREWORD

National security decisionmaking under stress or crisis management is something with which I have had some firsthand experience over the course of my career in government service. Most relevant to the topic of this edited volume is my tour of duty as U.S. Ambassador in Beijing which began in May 1989—a month before Tiananmen of June 3 and 4. In my position as chief U.S. diplomat in China, I was an actor and an observer—along with many dedicated and resourceful U.S. Embassy personnel—to the events that constituted a case study of Chinese communist crisis management. My colleagues and I were witnesses to what, in my judgment, constituted one of the gravest crises to the communists' control of China since 1949. We engaged the Chinese leadership during this time of tension and precipitous action.

As ambassador, I was responsible for managing the stressful and perilous situation that confronted the Embassy, its personnel, and U.S. citizens living and working in Beijing and elsewhere in China. While I directed our diplomatic personnel to do their utmost to report on and document the full extent of the crackdown ordered by China's communist leaders, my foremost concern was to ensure the safety and security of Americans in the country at the time. As a result more than 6,000 U.S. citizens were withdrawn from China in what was one of the largest evacuations of overseas Americans in a crisis situation since World War II. We saw both bravery and shirking among Americans, while the media was constantly trying to expose flaws in our actions. It was a time of rapid change and considerable manipulation.

As events such as the EP-3 incident of April 2001 and the severe acute respiratory syndrome (SARS) epidemic of 2003 demonstrate, the topic of this volume is as timely and important today as it was 16 years ago. An improved understanding of how the Chinese leadership—civilian and military—handles decisionmaking under conditions of crisis or stress is essential. This volume makes a worthwhile contribution on this topic. I commend it to you.

Ambassador James R. Lilley
Senior Fellow
American Enterprise Institute

CHAPTER 1
INTRODUCTION

Andrew Scobell and Larry M. Wortzel

If there is one constant in expert analyses of the history of modern China, it is the characterization of a country perpetually in the throes of crises. And in nearly all crises, the Chinese People's Liberation Army (PLA) has played an instrumental role. While China at the mid-point of the 21st century's first decade is arguably the most secure and stable it has been in more than a century, crises continue to emerge with apparent frequency. Consequently, the study of China's behavior in conditions of tension and stress, and particularly how the PLA is a factor in that behavior, is of considerable importance to policymakers and analysts around the world.

This volume represents the fruits of a conference held at the U.S. Army War College in September 2004 on the theme of "Chinese Crisis Management." One of the major debates that emerged among participants was whether all the case studies under examination constituted crises in the eyes of China's leaders. The consensus was that not all of these incidents were perceived as crises—a key case in point being the three Iraq wars (1980-88, 1990-91, and 2003). As a result, the rubric of "decisionmaking under stress" was adopted as presenters revised their papers for publication. No matter what rubric is employed, however, the chapters in this volume shed light on patterns of Chinese behavior in crisis-like situations and decisionmaking under stress.

Michael Swaine's contributed chapter first establishes a general framework for understanding crisis management based on previous work by Alexander George and J. Philip Rogers. He then proceeds to apply this framework to Chinese crisis management in particular. Swaine identifies five basic variables that influence crisis management behavior—subjective views of leaders and public, domestic environment, decisionmaking structure, information receipt and processing, and idiosyncratic features. In the case of China, he argues, the country often views itself as a victim and therefore strongly reacts to what it perceives as "unjust actions" on the part of other countries. Chinese leaders are thereby compelled to signal their firm resolve on serious issues through words and actions. However, decisionmaking is centralized in the hands of a small number of Party cadre, who work to develop a consensus, while China's bureaucratic Party and intelligence system severely compartmentalizes the flow of information, especially to senior leaders. This limits and sometimes distorts the information they receive during crisis situations.

Swaine then raises a number of questions about the factors that influence the Chinese framework for decisionmaking. He concludes that, if we can better understand the broad tendencies that affect China's crisis management style, we may be able to reduce the likelihood of undesirable situations in which a Sino-U.S. crisis would erupt.

Larry Wortzel presented a paper on Chinese decisionmaking and the Tiananmen Square Massacre. In Wortzel's opinion, at the time of Hu Yaobang's death, the Chinese Communist Party (CCP) was under intense public pressure to reform and reduce corruption. Hu's death acted as a catalyst, leading to student demonstrations, which were encouraged by reformist members of the CCP. As the protests became larger, several conservative factions, normally at odds with one another, closed ranks and sought to end the demonstrations, first through police, then military, means. However, the consensus decision to use force took time, and the apparent lack of action by the Party was seen by protestors to be tacit approval of their actions.

The People's Liberation Army (PLA) units that were finally sent in after the declaration of martial law were warned of "counter-revolutionary criminals" mixed in with the protesters. The residents of Beijing, who violently resisted the PLA in street battles, only confirmed the soldier's belief in these "criminals." The clearing of both the Square and Beijing were bloody, and shattered the myth of the PLA as "the people's army." Wortzel concludes that the CCP's style of "consensus leadership" does not lend itself well to crisis decisionmaking and that as long as it continues, it will likely hurt, rather than help, the management of future crises.

Susan Puska dissects the CCP's response to the severe acute respiratory syndrome (SARS) crisis in 2002-03. According to Puska, the CCP's top priority in the crisis was not public health, but party survival and maintenance of power. From November 2002 until April

2003, the Chinese government at first covered up the existence of SARS, then underreported the number of cases to the World Health Organization (WHO). Once motivated by public outcry, however, the government response was a massive, political-style anti-SARS campaign that did prove effective in stopping the spread of the disease.

Richard Bush's paper traces the influences and factors that both compose and complicate the PRC-Taiwan "question." In Dr. Bush's opinion, distrust on both sides of the Taiwan Strait, coupled with Taiwan's "de facto" alliance with the United States, creates a cycle wherein one side acts, the other side reacts—and sometimes over-reacts—and the United States is forced to deal with both parties.

Bush goes on to give an in-depth analysis of the 1995-96 Taiwan Strait crisis and discusses the factors that made up the problem. Among them are the decisionmaking processes of the highest levels of the PRC government, their strategic objectives, fundamental interests, problems in cross-strait communications, the lack of trust on both sides of the Strait, and U.S. fears of being sucked into an unwanted military action.

Paul Godwin, in his contribution, gives a detailed critical analysis of China's decisionmaking and negotiating strategy in both the 1999 Belgrade Embassy bombing and the EP-3 Surveillance Plane incidents. After describing how the two different events unfolded, Dr. Godwin points out the similarities in the Chinese handling of both. In both cases, China used a prolonged, "asymmetric" negotiating strategy designed to extract concessions from, and gain advantage over, the United States, while not causing serious damage or permanent rupture in Sino-U.S. relations.

Yitzhak Schichor analyzes the politico-military decisionmaking process of how China handled each of the "Iraqi" wars (Iran-Iraq, the First Gulf War, and the Coalition War in Iraq). The handling of the wars changed as China's role in international politics changed and grew, with China playing an increasingly important role on the world stage. But in all three wars, China benefited. In the Iran-Iraq War, China supplied arms to both sides and not only benefited economically, but was also able to see how their weapons systems worked under combat conditions. In the Gulf War, China used its potential veto and position on the United Nations (UN) Security Council to regain some of the international standing lost after the Tiananmen Square Massacre. And in both the Gulf War and the War in Iraq, China was able to see U.S. military technology in action, giving them useful information on a potential rival's military at no cost.

Lastly, Frank Miller and Andrew Scobell try to make sense of all the contributions. They ask whether "decisionmaking under stress" or "crisis management" are useful rubrics and argue that the latter remains a more useful approach and one that seems applicable to the case studies examined in this volume. Miller and Scobell contend that the Chinese Communist Party was born, nurtured, and matured in a climate of turmoil routinely punctuated by crises. China's communist leaders, the authors argue, find crises useful and even necessary to promote and preserve their rule. Miller and Scobell classify crises into three categories: fabricated, anticipated, and unanticipated. The first type is a crisis that has been manufactured by the regime to serve its purposes. The second type is a crisis that the regime sees coming and seeks to exploit to its advantage. The third type is one in which the regime is caught entirely by surprise and must scramble to respond. Beijing is entirely adept at the first instance, typically capable in the second instance, and at sea in the third instance.

What is of greatest importance in a particular situation is to identify which of these three types of crises (i.e., fabricated, anticipated, unanticipated) China is facing. Once this determination has been made, Chinese behavior becomes comprehensible and even predictable to an external actor who can then devise the appropriate response.

CHAPTER 2

CHINESE CRISIS MANAGEMENT:
FRAMEWORK FOR ANALYSIS, TENTATIVE
OBSERVATIONS, AND QUESTIONS FOR THE FUTURE

Michael D. Swaine

Numerous studies exist of China's use of force and its behavior in political-military crises. However, very few such studies examine Chinese views and actions from the perspective of crisis management. Yet the need for such an examination arguably has never been greater, given the occurrence of recent confrontations between the United States and China (such as the Taiwan Strait Crisis of 1995-96 and the EP-3 Incident of 2001), Beijing's increasing power and influence in Asia, and the arguably growing danger of a serious crisis emerging in the near to medium-term over volatile issues such as Taiwan, North Korea, and several territorial disputes along China's borders. To increase our chances of avoiding such crises in the future, or to minimize the damage that such crises might produce, we must greatly increase our understanding of the nature, scope, and requirements for successful crisis management; the specific conceptual and behavioral features of Chinese foreign crises that relate to crisis management; and the implications of such factors for U.S. crisis management behavior.

This chapter constitutes a first step in this direction. It is divided into four sections. The first section presents a general framework for understanding crisis management. The analysis identifies and/or defines basic concepts such as political-military crises, and the most significant crisis management strategies and bargaining approaches employed by nations in political-military crises. It also lists five basic sets of variables that most directly influence the crisis management behavior of individual nations. The second section identifies and summarizes the major features of Chinese crisis management behavior, with reference to the preceding conceptual framework. This summary is based on scholarly studies of the historical record and current evidence, largely in the form of recent interviews conducted by the author. The third section provides some preliminary observations relevant to the issue of whether China's crisis management behavior is likely to increase the chances of inadvertent escalation and/or conflict in a Sino-American crisis involving Taiwan. The fourth section identifies critical problem areas and unresolved issues concerning Beijing's likely current and future approach to managing political-military crises and draws some overall conclusions regarding the state of our knowledge about China's crisis management approach. Finally, two appendices list major sources on political-military crises and crisis management (Appendix I) and China's approach to conflict and crisis management (Appendix II).

SECTION 1. DEFINITIONS AND MAJOR VARIABLES

In Western analyses, a political-military crisis is usually defined by three factors:

1. A crisis involves the key or core interests of the actors involved;

2. There is a time element or sense of urgency after interests are invoked; and,

3. There is a possibility of greatly advancing and/or threatening the interests of both sides, including the threat of military conflict, and, in the case of major powers, a potential threat to the structure of the international system.[1]

An international crisis begins with a disruptive action or event, a breakpoint or trigger, which activates the above conditions for one or more states. Such a precipitating factor could occur by accident or deliberately; it could be entirely unexpected or emerge unsurprisingly (or seemingly unavoidably) from a longstanding, tense confrontation. It might also be caused by the actions of a third party or parties. In a full-blown political-military crisis, a threat of military conflict usually exists. In a near crisis, there is no probability of military conflict despite the existence of a conflict of interest and time pressures. Nonetheless, even near-crises can damage significantly the political, diplomatic, and economic relationships of the states concerned, and in some cases increase the probability of a future full-blown crisis.[2]

A true crisis emerges when neither side is willing to "back down" in the face of a perceived threat or opportunity. As Alexander George points out, some crises emerge in ways that leave one actor no choice but to counter its adversary; others emerge only because one actor decides to accept a challenge from the other and to oppose it; still other crises are deliberately initiated by one side in an effort to cause a favorable change in the status quo.[3] In other words, a crisis situation usually presents either an apparent threat or an opportunity (or both) for one or more states involved. A crisis does not necessarily include a threat or opportunity for both sides—only one side need perceive the existence of such factors for a crisis to occur.

A crisis ends with an action or event that denotes a qualitative reduction in conflict or greatly lowers the possibility of a conflict emerging. Thus, "successful" crisis management occurs when the parties involved are able to avoid the worst case and to defuse one or more elements of a crisis—particularly the possibility of military conflict—while also protecting or advancing their core interests.[4] The likelihood that any given crisis can be managed successfully will vary enormously, depending on many factors, including the skill and intent of the actors involved and whether the interests of the actors are diametrically opposed, quasi-opposed, or jointly opposed.[5]

In general, crisis management does not aim at resolving the basic issue or problem that created the crisis in the first place. It merely defuses the crisis and the risks of escalation.[6] As suggested above, crises differ substantially in their structure and dynamics, in the importance of what is at stake for all actors, in the larger diplomatic and military environment, in the level of risks and opportunities confronting each actor, and in the domestic and international constraints operating on key decisionmakers. However, the acute policy challenge posed by *every* political-military crisis emerges from the inherent tension that exists between the desire to protect or advance key interests and the need to avoid utilizing actions for this purpose that could bring about unwanted escalation and conflict. A political-military crisis management "bargaining" strategy is usually applied—whether consciously or unconsciously—to deal with this dilemma and to attain specific objectives.[7]

Crisis bargaining strategies vary in large part depending on the specific level and intensity of three basic components of bargaining—*persuasion* (including efforts to explain and justify a position and appropriate assurances of one's limited objectives), *accommodation* (including trade-offs, pay-offs, and the moderation of one's stand), and *coercion* (including everything from coercive signaling to an actual use of force short of all-out war)[8]—and the sequence in which these elements of bargaining are employed. Hence, many possible crisis management "bargaining" strategies exist, including both offensive strategies (i.e., compellence-oriented and intended to alter the situation at the expense of the adversary) and defensive strategies (i.e., deterrence-oriented and intended to prevent or reverse gains).[9] Moreover, more than one crisis bargaining strategy can be used in a crisis. Decisionmakers often differ greatly in their beliefs about these three different components of bargaining and the application related bargaining strategies. Such differences will influence how information is interpreted and the type of policy response favored during a crisis.[10]

Crisis management behavior does not consist simply of the application of bargaining strategies, however. It also is influenced heavily by a variety of other factors. J. Philip Rogers has developed a range of crisis bargaining "codes" or cognitive prisms that influence the way a decisionmaker interprets events and evaluates options in a crisis, thus influencing a state's overall approach to crisis management. These codes incorporate not only general leadership beliefs about the most optimal type(s) of bargaining strategy or strategies that should be applied in a crisis, but also two other core perceptions: the image of the adversary (comprising beliefs about the adversary's typical objectives, decisionmaking style, and bargaining strategy) and the dynamics of escalation, including the best ways to control escalation, and the manner in which a war might erupt in a crisis. Indeed, in Rogers's analysis, the latter two factors greatly influence the type of bargaining strategy or strategies adopted by a state actor in a crisis.[11]

Rogers identifies four basic bargaining codes, each representing an ideal type. Type A employs an image of the adversary that almost entirely is aggressive in its goals. Adherents believe that only intentional war is possible; that crisis escalation is easily controllable even beyond the onset of limited nuclear exchanges (and certainly controllable through very high levels of conventional combat); and that crises can escalate to war only when leaders on one side view the balance of force as favorable or when they view the other side as lacking resolve. Hence, crisis management is focused on efforts to anticipate and avoid both gaps in one's military or intelligence capabilities and an inadequate demonstration of one's willingness to use force. Because of their belief in the low likelihood of unwanted escalation, proponents of this crisis management style usually adopt a warfighting approach to deterrence in a crisis. That is, when leaders calculate that there is a fairly high chance of military success, they will reject coercive diplomacy or other forms of crisis bargaining strategies in favor of a preemptive military fait accompli. If chances of success are not deemed high or are uncertain, they will opt for strong, dramatic, coercive actions. In addition, this approach also tends to define objectives and evaluate success or failure in military terms and to downplay potential negative political costs in a crisis. Adherents of this approach often employ bluffs and can use nuclear threats.[12]

Type B is similar to Type A, but differs notably in the significance it accords to the political context of a crisis and in the image of crisis dynamics. Adherents admit there is a brink or point beyond which crisis control will become problematic and that one can lose control of the escalation process, even before actual conflict occurs. An unintended escalatory spiral can cause one side to believe an attack is imminent and proceed to launch a preemptive strike. However, adherents of this approach believe they have a pretty clear understanding of the dynamics of escalation and hence of actions that could lead to unwanted escalation. They are thus more receptive to the use of coercive diplomacy than Types C and D below. But they also believe that a loss of control can result from technical breakdowns and the path that strategic interaction can take. Hence, they will avoid actions that could in their view prompt inadvertent escalation. Among adherents, two viewpoints exist regarding what might produce such escalation. One variant is willing to use both bluffs and nuclear threats. They believe that nuclear alerts can be safer than conventional threats and attempts at incremental escalation, because of the danger that tit-for-tat exchanges can lead to a loss of control, and because conventional combat is more likely to lead to inadvertent war due to the difficulties involved in controlling conventional forces. They emphasize the need to convey sufficient resolve early in the game to avoid a loss of control and believe that nuclear bluffs and alerts can convey this most effectively. A second variant believes that bluffing, and nuclear bluffing in particular, is difficult to accomplish and dangerous. Hence, they are more willing to use conventional forces to achieve a non-nuclear fait accompli. However, both variants believe that dramatic escalations often are safer than incremental ones. They both believe that small steps will be seen as timid and invite counterescalation. For adherents of this approach, the most common cause of war in a crisis is due to a failure to demonstrate resolve early and dramatically. Both variants thus have fairly high confidence that inadvertent escalation can be avoided by following their formulas or rules. They may also be more inclined than Type A to include carrots in crisis bargaining. Overall, the problem this type poses for crisis management is its emphasis on relatively strong, coercive actions in the initial phases of a crisis.[13]

Type C believes it is often difficult to determine whether the objectives of the adversary in a particular crisis are mainly offensive or defensive. The interpretation of an adversary's objective is more influenced by the situational context. Adherents of this approach hold two images of the causes of war: 1) a failure to demonstrate resolve; and 2) spiraling escalation and responses to perceived provocations. Both of these features provide ample opportunities for mistakes and hence many paths to inadvertent war. This type is thus less sanguine about controlling crises, especially via dramatic escalation. Adherents believe that strong signals could set the escalatory process in motion, and hence they do not agree with Type B that the brink can be recognized in advance and that one can control events right up to the brink. Hence, adherents of this approach will favor incremental coercive escalation over more sudden and dramatic forms of escalation. They hold a strong aversion to fait accompli strategies, because of the great uncertainties inherent in the two possible images of crisis escalation mentioned above. The incremental approach offers a compromise between the two possible images of the adversary's objectives and of crisis dynamics. An incremental strategy can show resolve but with less provocation, and stress the use of mixed (carrot and stick) elements. This facilitates efforts to steer between inadvertent war and appeasement. Adherents of this approach are thus more willing to compromise on nonessential issues and more sensitive to situational variables.[14]

Type D tends to assume that the adversary operates from exclusively defensive motives. Hence, control is extremely problematic if one puts even a modest stress on coercion. For adherents of this approach, war results from the escalatory spiral resulting from coercive moves, not from a failure to show resolve. Thus, this approach stresses accommodation and crisis prevention, not management or bargaining.[15]

On the basis of both historical experience and (in particular) rationality-based calculations of risk assessment, Western analyses of crisis management behavior suggest that some of the above strategies, bargaining codes, and approaches are far more likely than others to decrease (or increase) the chances of inadvertent escalation and conflict in a political-military crisis, and to protect (or weaken) a state's core interests. This is largely because each strategy is more or less likely to facilitate the use of what are regarded by many scholars as several prudent political and operational requirements for "successful" crisis management (i.e., so-called "rules of prudence").[16] Political requirements include the use of limited objectives and limited means on behalf of such objectives, and an avoidance of the use of ultimatums.[17] They also include efforts to avoid "ideological" or "principled" positions that encourage "zero-sum approaches to a crisis and might threaten the other side's core values and mix moral principles with conflicts of interest. Operational requirements include the need to preserve military flexibility, to escalate slowly, to avoid excessive pressure, to exercise self-restraint, to communicate clearly and consistently and to be specific about demands.[18]

As Alexander George states, from this perspective, of the five offensive strategies, the least dangerous or risky are the limited probe and controlled pressure, because they give a challenger a good opportunity to monitor and control risks. In contrast, the blackmail and fait accompli strategies are based on the assumption that the defender will be too intimidated or insufficiently motivated to resist, or that he will not respond with military action because he has made no prior commitment. If such assumptions are incorrect, war might follow rapidly as these strategies allow little opportunity to monitor and control risks. Finally, a strategy of slow attrition might entail low risks at first, but could force the defender to escalate greatly, according to George, as the defender is bled to the point where he is provoked to undertake a major provocation.[19]

Of the seven defensive strategies, George argues that the least risky among them, at least at the beginning of a crisis, are the tit-for-tat, the test of capabilities, the drawing a line, efforts to convey commitment and resolve, and strategies centered on various time-buying efforts. More immediately and significantly risky are strategies of coercive diplomacy and limited escalation. The latter strategy arguably only works when paired with effective deterrence of counterescalation by the adversary. George states that "... coercive diplomacy is a particularly beguiling strategy for strong powers that suffer an encroachment from a weaker state because it seems to promise success without bloodshed or much expenditure of resources." However, he argues that proponents of this strategy often fail to consider whether a weaker opponent's strong motivation might compensate for its inferior capabilities and thus lead it to counter vigorously attempts at coercion. This might be particularly applicable in the case of a U.S.-China crisis. This strategy also is highly problematic if it is combined with stringent demands that strengthen the opponent's motivation to resist.[20]

While low risk strategies clearly offer some benefits by reducing the possibility of conflict, George also points out that they might also produce serious disadvantages in a crisis. For example, an exclusive commitment to accommodationist, low risk strategies might ultimately fail by preserving peace at the expense of core state interests. Also, such strategies might prove to be entirely ineffective or, worse yet, to convey weakness to the opponent, thus emboldening him to use coercion or force. Finally, the effectiveness of various crisis management strategies can also be greatly influenced by the process of implementation. For example, strategies that look great on paper might fail due to bureaucratic obstacles, or a heavy reliance on subtle signaling or unusually timely or reliable communication between opponents.[21]

Overall, assessments of the benefits and risks of the above crisis management strategies suggest that the bargaining "code" or paradigm identified as Type C above would likely offer the greatest chance of "success" from a purely logical or conceptual viewpoint. At the same time, the Type C code is not without its problems. As Rogers points out, adherents of this approach are arguably more likely to procrastinate, which can cause a serious problem when the crisis demands courageous and bold leadership. They also are likely to experience significant levels of stress, which can reduce the quality of decisions. Moreover, practitioners of this approach might display a greater degree of fluctuation in behavior (especially between tough and conciliatory responses), which can send confusing signals to the adversary. While Rogers believes that Types A and B pose an excessively high level of risk in a crisis, the extremely cautious Type D code is also dangerous, for reasons discussed in the previous paragraph. The Type D code is dangerous particularly when the adversary is highly aggressive and has strong ambitions to alter the status quo. In general, however, Rogers believes that the Type A crisis bargaining code is probably the most dangerous, since adherents of such an approach tend to employ strong, coercive actions or undertake preemptive military moves even when they hold limited political objectives. This can make an opponent believe that its basic security interests are threatened.[22]

The Rogers typology of crisis bargaining codes and those constituent strategies enunciated by George together provide a useful basis for defining the basic "ideal" alternative cognitive approaches to crisis management held by decisionmakers in a political-military crisis. However, at best, this schema only identifies distinctive subjective *tendencies* that individual decisionmakers might display in their efforts to manage such a crisis. As Alexander George states, actual crisis management behavior is highly context-dependent, i.e., "... the prospects for and outcomes of crisis management are subject to the interplay of many variables that are likely to present themselves

in somewhat different configurations in different confrontations."[23]
George asserts, for example, that "success" in crisis management
(as defined above) is highly dependent on: 1) the strength of the
decisionmakers' incentives for avoiding war; 2) the opportunities
available to decisionmakers for managing crises; and 3) the level of
skill they bring to bear in any crisis management effort.[24] Moreover,
Rogers points out that crisis bargaining codes or approaches will
operate somewhat differently when the decisionmaker is faced
with different information problems, i.e., information overload,
information deficits, and contradictory information. Crisis
management behavior also will operate differently as a result of other
factors that Rogers mentions only in passing, such as the larger self-
image leaders hold of their country and its core values; the features of
the domestic political, social, and economic environment confronting
decisionmakers; the characteristics of the decisionmaking process;
and various idiosyncratic factors such as leadership personality,
climate and weather, technical issues, the effects of stress, etc.[25]

In all, one can identify at least five basic sets of variables that
influence crisis management behavior:

1. *Subjective Views/Beliefs of Leaders and the Public.* This set includes
   elite attitudes toward risk taking and crisis stability; the self-
   image of the populace as a nation and a people; images of
   the adversary; fundamental values and assumptions held
   regarding coercion (and the use of force), accommodation,
   and persuasion, and hence the basic inclination toward
   specific bargaining strategies; and, finally, views of the best
   means to signal credibility and to resolve or avoid unwanted
   escalation.

2. *Domestic Environment (politics and society).* This set includes
   the system of government and the nature of leadership
   politics, the elite's perceived requirements for governmental
   and social stability and progress (including governmental
   approaches to manipulating beliefs and images, the media,
   etc.), and the nature and level of influence of public opinion
   and other forms of popular pressure on the government.

3. *Decisionmaking Structure and Process.* This set includes the
   pattern of distribution and exercise of ultimate decisionmaking

power within the political system, as well as both formal and informal channels of policy execution and control; this area thus encompasses bureaucratic relations and interests (especially the relationship between civilian and military leaders and organs), relationships between political and military leaders and their subordinates, the speed and efficiency of the decisionmaking structure, and the influence of standard operating procedures (SOP) and signaling.

4. *Information and Intelligence Receipt and Processing.* This set includes the nature and extent of governmental intelligence gathering and use by each actor, as well as the acquisition and use of other forms of information, etc.; it encompasses the possible influence exerted by information overload, information deficit, and contradictory or distorted information, as well as other features of information input and utilization by crisis decisionmakers and subordinate individuals and agencies charged with implementation.

5. *Idiosyncratic or Special Features.* This set includes all those irregular or unpredictable factors that can influence crisis management, such as leadership personality, the effects of stress and climate, technical issues or problems, and the effect of third parties.[26]

In sum, every state's crisis management approach, both in general and during a particular crisis, can be described as an approximation of a certain type of crisis bargaining "code" or cognitive schema that incorporates specific types of crisis management strategies. Moreover, each such code or schema is influenced or qualified by exogenous factors such as domestic pressures, decisionmaking structures and processes, intelligence features, the number of state actors in a crisis, etc.

SECTION 2. CHINA'S APPROACH TO CRISES AND CRISIS MANAGEMENT

In order to determine the major characteristics of China's crisis management approach as it relates to the above framework, one must first identify those dominant Chinese beliefs, values, and actions in the above five component areas that exert the most direct influence on crisis behavior.[27]

Subjective Views/Beliefs.

Self-Image. China views itself as an aspiring yet nonaggressive great power, increasingly confident yet also acutely sensitive to domestic and external challenges to its stability and status. China's leaders, and many ordinary Chinese citizens, possess a strong memory of the nation's supposed historical victimization and manipulation at the hands of stronger powers. Thus, they are prepared to go to significant lengths to avoid the appearance of being weak and "giving-in" to great power pressures, or of engaging in predatory or manipulative behavior themselves. Chinese leaders also evince a very strong commitment to specific basic principles and core interests, especially those principles and interests associated with the defense of China's territorial integrity and sovereignty, both of which are related closely to national dignity. This viewpoint is apparently also shared by many ordinary Chinese citizens.

China generally has viewed its behavior during most past crises over territorial issues as a totally justifiable defensive response to efforts by other states to alter the territorial status quo accepted by China's leaders in 1949. That is, the use of force in most post-1949 territorial crises is viewed as a defense against threats to the status quo, i.e., a kind of "preventive deterrence action" designed to prevent the situation from worsening. China's involvement with foreign powers in other political-military crises since 1949 also is characterized as largely defensive in nature, designed to ward off either imminent or existing threats to critical border areas, or more vague attempts to intimidate China or to "test" China's resolve or the stability of its leadership.

Closely related to the previous point, China also displays a strong impulse to view the triggering issue in a crisis as a clear matter of principle (i.e., of right and wrong, fairness or unfairness, or just versus unjust behavior). This leads to a tendency to view crisis confrontations in "zero-sum" terms, involving the defense of moral principles against unjust acts. This tendency is augmented by the above-mentioned sense of vulnerability felt when confronting a superior power. China thus often believes it is compelled to act because the other side would not heed warnings and recognize its unjust behavior, or because the other side acted in a way that equated to an unjustifiable use of force, requiring a counter.

The prevalence of territorial and sovereignty issues in past crises, and the relevance of such issues to the current Taiwan problem, reinforce a strong Chinese belief that Beijing is very likely to have its most critical interests at stake in a future crisis with the United States over the island. At the same time, many Chinese also assume that Beijing also will be defending such interests from an inferior position in power terms. In contrast, the United States is viewed as likely to have lesser interests at stake in a Taiwan crisis.

Image of the Adversary (the United States). China views the United States as constantly striving to maintain its system of global and regional dominance, usually through a reliance on superior economic and military power, and often without international (i.e., United Nations [UN]) approval. In particular, Washington is seen as willing to violate the territorial integrity and sovereignty of other states to achieve its objectives. Moreover, as a hegemonic and anti-socialist power, the United States inevitably views China as a significant strategic threat. Hence, the United States often is seen as offensively-oriented, seeking in many ways to constrain China's power and limit its options internationally. In the past, these efforts have included attempts to use other powers, such as Nationalist China, South Korea, and Japan as proxies. Today, the United States often is viewed as seeking to constrain China's rise by preventing the reunification of Taiwan with the Mainland, or of encouraging Taiwan independence. On the other hand, the United States also is viewed today as desiring at least workable (if not fully cooperative and amicable) relations with China, for largely economic and political reasons. This U.S. interest has deepened considerably since the advent of the global war on terrorism and the worsening of the current crisis on the Korean Peninsula. Both events have forced Washington not only to divert its attention, at least temporarily, from the long-term strategic challenge posed by China's rise, but also to collaborate more closely with and depend upon Beijing to address these more pressing concerns.

The United States also is viewed as possessing a much larger variety of means to manage a crisis, including military, political, economic, diplomatic, etc. On the other hand, it also will move slowly in a crisis and usually is constrained by a fear of casualties, prolonged conflict, and economic costs. It also displays some clear vulnerabilities in command, control, and communications, stemming largely from the U.S. military's reliance on nonencrypted electronic communications and satellite technologies. Moreover, many Chinese firmly believe that the United States probably will have less at stake in a conflict with China, as suggested above. For all these reasons, many Chinese apparently believe that the United States can, in most instances, more easily choose to avoid a crisis with China involving the use of force over critical issues such as Taiwan, viewing armed conflict—and particularly the prospect of prolonged armed conflict—as unnecessary and too costly. This suggests, to some Chinese, that the United States is likely to be more easily deterrable in a crisis than China, especially a crisis over territories such as Taiwan.

Also, China often places an emphasis on designating an adversary as friendly, hostile, or neither. In general, China is more inclined to adopt an enemy image of an adversary if overall bilateral relations are in a state of hostility or obvious tension, or simply if friendly relations are not predominant in the relationship (i.e., a state of "neither friend nor foe," *fei di fei you*). Such a designation is apparently more than an informal, subjective opinion held by some leaders. It is a quasi-formal "label" (referred to by the notion of *dingwei*) that can heavily influence assessments and recommendations produced by elites and advisors within China, and thus can significantly shape Beijing's crisis behavior.

Views toward Coercion, Accommodation, and Persuasion. Historically, Chinese behavior in political-military crises has encompassed at times all forms of coercion, including the direct application of military force. Indeed, for China, a limited use of force has been regarded as an effective tool in a crisis. Such a use of force can be used to show resolve, a commitment to principle, and a refusal to submit to intimidation, and thus can elicit caution and possibly concessions from the other side. It can be designed to produce psychological shock, uncertainty, and to intimidate an opponent, often as part of a larger strategy to seize the political and military initiative via deception and surprise. From the Chinese perspective, such a limited use of force under certain circumstances can prevent a much larger conflict. In all, in past political-military crises, China's use of force often was intended to shape, deter, blunt, or reverse a situation; probe or test intentions; prevent escalation; and, in the Chinese view, strengthen the foundations of peace.

In some instances, a self-perception by China of overall weakness, not strength, can motivate the use of force as a means of conveying resolve and to shock a stronger adversary into more "cautious" behavior. Such a use of force usually demands sensitivity to the prevailing balance of power in the geographic area of the crisis and to problems of escalation and control. In line with this approach, in past crises, the Chinese use of force often was followed by signs of accommodation or efforts at persuasion, at least privately, to avoid escalation, and to secure at least minimum gains.

Overall, Chinese leaders have seemed to follow the maxim "just grounds, to our advantage, with restraint" (*you li, you li, you jie*) in assessing how to employ coercion, accommodation, and persuasion in a crisis bargaining strategy. This principle, used often by Mao Zedong during the Chinese struggle against Japan in World War II, consists of three points:

1. Do not attack unless attacked. Never attack others without provocation, but once attacked, do not fail to return the blow. This is the defensive nature of the principle.

2. Do not fight decisive actions unless sure of victory. Never fight without certainty of success, unless failing to fight would likely present a worse outcome. Utilize contradictions among the enemy. Apply your strong point(s) and reduce the enemy's strong point(s). This is the limited nature of struggle.

3. Be pragmatic and aware of the limited nature of objectives and strength. With a strong power, set appropriate war objectives; do not exceed capabilities. Know when to stop, when to counter, and when to bring the fight to a close. Stop once the goals are attained; rethink if you cannot obtain your objectives. Do not be carried away with success. This is the temporary or contingent nature of each struggle.

Views toward Escalation. In past crises, the above approach to force usually required the prior attainment of local superiority, strong control over troops (marked by very clear rules of engagement), efforts to seize and maintain the initiative (often using tactical surprise and deception), a sense of "knowing when to stop," the use of pauses, and/or the communication of what were viewed as clear signals of a low intent to escalate in a major way (e.g., no obvious alerts or mobilizations, etc.). In most instances, the provision of a "way out" for both sides also was emphasized. Such notions are broadly similar to the "rules of prudence" contained in the general literature on crisis management. When combined with attempts to maximize constraining influences on the adversary (such as via messages aimed at world and domestic public opinion, etc.), Chinese leaders believed that these conditions would minimize the likelihood of miscalculation or of a preemptive attack by the adversary, and thus limit escalation. This would be especially true for those crises involving the use of force to attain limited, primarily political, objectives. Such a viewpoint applied even against a superior (including a nuclear-armed) foe, particularly if vital interests were at stake for China and if delay was seen as more dangerous than action. From the Chinese perspective, successful examples of the application of this approach include the Taiwan Strait crises of 1954 and 1958 and the 1962 Sino-Indian border clash.

For Chinese leaders, signaling firm resolve through words and actions also is a major element of crisis bargaining and escalation control. In past political-military crises during the Mao Zedong and Deng Xiaoping eras, Chinese leaders showed a clear willingness to sustain significant military and/or economic costs, if they were confident that, by doing so, they could attain their core (usually political) objectives. From a Western viewpoint, this amounts to risking significant escalation and subsequent damage for limited objectives. From the Chinese perspective, the willingness to put major assets at risk in a crisis is an essential means of signaling resolve. Moreover, as indicated above, Chinese leaders tend to believe that a strong show of resolve is needed in part to compensate for relative weakness. And it is used not only to deter, but also to justify subsequent actions, both externally and domestically.

In confronting or planning for a crisis, Chinese leaders also at times have emphasized the need to achieve the best outcome possible while preparing for the worst. The Chinese apparently assume that, in order to control escalation and manage a crisis successfully, China must be prepared to deal with—indeed to anticipate—the most severe deterioration of the situation possible. To some extent, this constitutes prudent planning. However, some Chinese observers have pointed out that an overemphasis on "worst casing" a potential crisis can serve to magnify the threat posed by an opponent, thus raising the cost of implementing any potential crisis–inducing policy to excessively high levels. On the other hand, such careful preparation for or consideration of worst case outcomes cannot occur if a crisis emerges unexpectedly and rapidly.[28]

Views toward and Features of Crisis Signaling. Historically, China has evinced little, if any, deliberate use of tactical ambiguity in signaling, especially in signaling resolve. In general, Chinese leaders seem to value highly sending what they view as clear messages. Such signals usually are intended to convey warnings and thus to deter an adversary, rather than to negotiate the resolution of a crisis, to indicate a willingness to deescalate, or to avoid further escalation. In the past, China has warned adversaries to alter their behavior or suffer the consequences, often as part of a prior internal decision to use force if China's warnings are not heeded. In other words, there is considerable evidence that China employs signals during a crisis primarily to communicate resolve and resistance, and not to attain the objectives usually associated with "classic" crisis management signaling.[29]

Moreover, China seems to emphasize verbal warnings in political-military crises, while resisting the use of overt military deployments or alerts. This presumably reflects the desire to employ surprise and to concentrate superior force at a point of weakness, especially if the adversary is regarded as militarily superior. In contrast, the United States arguably views most forms of military transparency in a crisis as a way to enhance deterrence.

Despite an emphasis on clear signaling to convey resolve, China's messages have not been interpreted as intended in some past crises, partly because other signals seem to contradict or weaken the intended message, sometimes for inadvertent reasons. For example, to some U.S. observers such as Tom Christensen, during the Korean War, China's warnings to the United States concerning the crossing of the 38th parallel were not clear particularly because of the lack of direct contact between the two sides, and because the Chinese did not use military deployments or alerts to signal resolve. In fact, the lack of coordination in China's military deployments before the war and in the early phases of the war sent signals of weakness or irresoluteness that were not real. Today, Chinese signals can be misinterpreted because the Chinese system is less monolithic than in the past, and somewhat different messages can emerge from different individuals and organizations. During the Maoist era, strong centralized control over many, if not all, aspects of crisis decisionmaking usually guaranteed a single message. Today a much more complex decisionmaking process exists, marked by far higher levels of internal consultation; as a result, different messages can emerge during a crisis. This can slow down reaction time and distort signaling.

Finally, as with other countries, both the sending and the reading of signals during a crisis are influenced heavily by the larger external and internal political context. For example, China's leadership tended to worst case U.S. signals during the 1995-96 Taiwan Strait crisis because bilateral relations had worsened considerably prior to that event. This negative environment led the Chinese leadership to conclude that Washington was probing China's bottom line when it: a) reversed its stance and granted a visa to Lee Teng-hui; b) allegedly refused to work with Beijing to lessen the consequences of the decision; and then c) deployed two carrier battle groups to the Taiwan area. Actually, according to former senior U.S. officials involved in the crisis, domestic concerns unrelated to China (in particular, a fear of growing congressional influence over the president's foreign policy authority) apparently played a major role in these decisions and signals. Similarly, the military signals Beijing employed in fall 1995 and spring 1996 at least were determined partly by domestic considerations, including both the desire to influence Taiwan's presidential election and the desire to mollify internal hardliners.[30]

The Domestic Environment.

For China, domestic factors often are critically important in a political-military crisis, and, in some instances, arguably more important than external factors. In fact, some Chinese observers insist that domestic interests always will trump foreign policy interests in a crisis. Both elite and popular views and actions involving internal issues and concerns can limit options, increase rigidity, slow response times, and distort signals in a crisis. Although the general contours of how domestic factors such as internal power disputes or elite concerns over popular sentiments might influence crises are understood, little is known about the critical details regarding the specific manner, degree, and conditionality of such influence. Individual anecdotes abound concerning specific cases, but general principles or features are virtually nonexistent. Moreover, overall, some Chinese observers believe that Beijing often displays a woefully inadequate understanding of how an adversary's domestic context influences its crisis behavior.

In the Mao Zedong and Deng Xiaoping eras, crises apparently were regarded at times as opportunities, including opportunities to strengthen a leader's domestic standing among his colleagues and with the public. And public opinion was regarded more as an element that should be mobilized to support the elite's objectives than a factor that conditioned crisis decisionmaking. For many Chinese observers, the Korean and Vietnam Wars were mostly viewed by the Chinese leadership as international crises without significant internal components. That is, Chinese leaders generally were not subject to serious domestic (and in particular public) pressures when handling those early crises. This is no longer the case, however. For example, during the diplomatic crises sparked by the U.S. bombing of China's embassy in Belgrade in 1999 and the EP-3 incident of 2001, China's leaders faced the problem of how to manage a situation in which internal dissatisfaction, sparked by an external crisis, threatened to imperil social stability. In fact, as suggested above, some Chinese observers believe that public pressure has become a far more serious issue for China's leaders than for America's leaders. Moreover, according to some Chinese observers, during the embassy bombing crisis of 1999, the influence of domestic factors clearly outweighed external considerations. That said, we still know very little about the ways in which public opinion or pressure can influence the calculations of Chinese leaders in a crisis.

In addition to public pressure, other domestic elements within China, such as the media and the propaganda apparatus, also play an important role in influencing a crisis. This is an area that requires much further study. From the perspective of some Chinese observers, the People's Republic of China (PRC) propaganda apparatus and the media are not entirely monolithic entities subservient to the views and decisions of the senior leadership. The propaganda apparatus tends to be more conservative than other institutions (and perhaps some leaders) and primarily is oriented toward domestic audiences. The broader media are more diverse, and dominated by younger individuals and sometimes both direct and reflect public sentiment. However, according to some Chinese observers, after the embassy bombing incident of 1999, the Chinese government greatly strengthened its control over the media, especially the mainstream media. The media reportedly now are included in consultative mechanisms and within the Chinese government's coordination mechanism.

Finally, since the Mao and Deng eras, groupings within the senior Chinese leadership also have come to play an increasingly important role during crises. In particular, the existence of a more collective leadership and differences between "hard-line" and "soft-line" approaches to handling a crisis (both within the public and among the elite) can influence outcomes. According to some Chinese, China's top leaders especially need to explain and justify themselves to those forces internally advocating a tougher line against Taiwan, for example. But, again, we understand very little about how such leadership differences might influence actual behavior in a crisis.

The Decisionmaking Structure and Process.

During the 1950s, 1960s, and early 1970s, the decisionmaking process during a crisis was highly centralized and concentrated in the person of Mao Zedong. At such times, Mao often consulted with a small group of senior civilian and military leaders, and at times would exert considerable efforts to persuade some of these individuals to support his viewpoint. The ideal procedure followed was democratic centralism, in which all senior leaders would exchange their views through high-level meetings and less formal interactions and a general consensus would eventually be reached on an issue, thus leading to a decision. However, in most instances, the resulting "consensus" generally would reflect Mao's preference, and in most instances Mao was able to make key decisions regardless of the level of hesitation or even opposition that might exist among his colleagues. In addition, Mao was able to make the most authoritative strategic and tactical decisions without much interference from the bureaucracy. Moreover, lower-level officials were almost completely in the dark regarding a particular decision, yet would implement the decision voluntarily, and follow established policies. Hence, a single decision with a single message usually emerged from the Mao era crisis decisionmaking process. Moreover, Mao also would make all critical decisions throughout the life of the crisis; for example, in response to the reaction of the adversary. At the same time, Mao's superior authority would allow him to strike compromises in a crisis that less powerful leaders might be unwilling or unable to make, for fear of being attacked by their rivals. These features also were evident by-and-large during the Deng Xiaoping era of the mid/late-1970s, 1980s, and early 1990s. However, Deng arguably was more compelled than Mao to consult and compromise with his senior colleagues, especially those retired or semi-retired cadres of the revolutionary generation.

The present-day Chinese crisis decisionmaking process remains highly centralized in a small number of individuals, but is not as fully concentrated in the hands of the paramount leader. The process is much more amorphous, involving greater levels of internal (and some external) consultation and a genuine need to develop a consensus through democratic centralism. The paramount leader still makes the final decision, but his decision must more closely reflect the views of his colleagues. In general, rising nationalist feelings within the public and the greater complexity of some issues confronted in a crisis require much greater levels of coordination within the government, and at times the solicitation of views from nongovernment officials such as scholars. Moreover, more diverse messages emerge from the Chinese government during a crisis today, including from different bureaucracies and the media. This reflects the involvement in various aspects of the decisionmaking or implementation process of a larger number of more *relatively* autonomous actors than existed during the Mao and Deng eras. This more complex and diverse process tends to slow down reaction time and distort signaling, as indicated above. In addition, whereas Mao or Deng might have been more willing to take risks in managing a crisis, on the assumption that their dominant power and authority made them less susceptible to removal if they were to fail, today's relatively weaker leaders might be more vulnerable to public or elite pressure and hence more cautious in a crisis.

One major bureaucratic player in the crisis decisionmaking process is the Chinese military. During the Maoist era, senior military leaders enjoyed considerable prestige and were well known by their colleagues within the civilian apparatus, many of whom were former military leaders. Hence, individual senior officers could and sometimes did advocate their views with vigor and even question Mao's viewpoint at times. If the decision in question involved military deployments, Mao would have to listen carefully to their views. Yet the military did not in any sense "check" Mao's decisionmaking power. Moreover, even though military commanders would have considerable freedom in implementing the orders given to units in the field, Mao would usually issue all such orders, be informed continuously of the movement of all major units, and at times even would direct personally the movement of units. He also would establish and ensure the observance of strict rules of engagement by all military forces deployed during a crisis.

Today, military leaders do not have close personal ties to civilian leaders. The relationship between senior civilian and military heads is largely professional, mediated and shaped by the functions and responsibilities of their respective institutions. Some personal links apparently do form at senior levels as a result of the personnel promotion process and frequent contact during policy meetings. However, the high level of personal familiarity between senior civilian and military leaders, the resulting close interactions among them, and the significant authority and power of senior military leaders within upper decisionmaking circles evident during the Maoist era in general do not exist today.

The views of military leaders during a crisis must be taken into account if relevant, but, as in the Maoist era, they do not translate into an absolute veto over decisions made by civilian leaders. That said, information supplied to the senior leadership by military sources, together with the operational plans and procedures of the military, can shape significantly the perceptions of senior civilian decisionmakers and hence the ultimate behavior of the leadership in a crisis. Such influence might have taken place during the EP-3 incident and the 1995-96 Taiwan Strait crisis. Military reports on the aircraft collision during the former were apparently the sole source of information on the incident, and in all likelihood could not be confirmed independently by the civilian leadership. During the latter crisis, according to Chinese observers, a preapproved operational plan drawn up by the military for exercises and missile firings apparently was never reexamined as the crisis evolved, and thus likely influenced the outcome of the crisis.

Finally, overall, the decisionmaking process can affect significantly information processing, signaling, and hence the willingness or ability of the Chinese leadership to adopt particular crisis bargaining strategies. During the Belgrade embassy bombing incident, the Chinese response was viewed by many observers as too slow. Why did the Chinese government wait to conduct a direct dialogue with the United States after the bombing? Was this solely because no authoritative response had been developed, or were other factors at work? Could some level of direct contact occur even in the absence of a formal decision defining China's official response? According to some Chinese observers, special task forces reportedly have been formed during recent crises to handle these and other problems. But few details are known about their membership, responsibilities, functions, and ultimate influence.

Information and Intelligence Receipt and Processing.

The receipt of a wide range of accurate, real-time intelligence and information (including so-called "soft" intelligence), incorporating, when necessary, a variety of perspectives and interpretations, is essential to effective crisis management. National policymakers' control over and access to both hard and soft intelligence has a direct impact on the quality of the final judgment. At the public level, this also influences the environment for determining security policy. Unfortunately, although the amount and quality of the intelligence and information received by China's leadership during crises reportedly has improved significantly over the years, problems remain, and much is still unknown about how such vital input is utilized.

For example, according to some Chinese observers, communication links with senior leaders do not always operate quickly and effectively, as indicated during the EP-3 incident. At that time, senior leaders reportedly could not be informed quickly of the incident because they were engaged in tree planting in various places around the country. In addition, the system for providing intelligence apparently reflects the overall "stove-piped" nature of the Chinese bureaucracy. For example, entirely separate avenues of intelligence exist in the civilian and military realms. Although this also is the case in other countries, in China, the military intelligence apparatus, in particular, remains insular, secretive, and highly restricted, even to some senior civilian leaders. This phenomenon arguably can limit or distort the information provided to senior decisionmakers in a crisis. It is possible that this occurred during the EP-3 incident. In general, this critical set of variables influencing China's crisis management behavior is very poorly understood in the West, and even, it seems, within much of the Chinese political system.

Idiosyncratic or Special Features.

One of the most important factors influencing the past management of Sino-U.S. crises has been the personality of China's senior leadership. According to most observers, crises such as the Korean War and the Sino-Soviet border conflict of 1969 were influenced decisively by the outlook and values of Mao Zedong. That said, it remains extremely difficult to distinguish between Mao's power and the unique features of his personality as determinative factors in the crisis decisionmaking process. Was Mao, as a personality, more willing to take risks, less "rational," and more rigid and resistant to negotiation in a crisis? Even more important, little evidence exists to determine the extent to which the personalities of today's leadership in China might affect their approach to political-military crises.

Another critical factor that can influence crisis decisionmaking is the activities of third parties. For example, the policies of Chiang Kai-shek significantly influenced the handling of the Taiwan Strait crises of the 1950s, sometimes increasing the danger involved. Taiwan's domestic politics and the personality of individuals such as Lee Teng-hui also have exerted a major impact on more recent crises. During the 1995-96 Taiwan Strait crisis, Lee Teng-hui had his own agenda, which put the United States in the position of needing to respond to his actions, as well as those of the Chinese. This is arguably also the case with Chen Shui-bian today. In other crises involving China, such as the 1962 Sino-Indian border crisis and the 1979 Sino-Vietnam border war, the policies and actions of the Soviet Union obviously exerted an influence over Chinese behavior toward New Delhi and Hanoi. The manner and extent to which third parties have influenced China's crisis management behavior requires much further study.

What tentative conclusions can one draw from the above features concerning China's past approach to crisis management? First, Beijing generally has been pragmatic and attentive to relative capabilities and the dangers of escalation and miscalculation in a crisis. As a result, it has displayed at least some important "rules of prudence," including a strict control over military forces, the use and signaling of limited objectives and means, and the provision of pauses and a "way out," when threatening or employing force. On the other hand, China's leadership also has displayed a clear tendency to regard an adversary as aggressive and often as an "enemy," and generally has utilized zero-sum thinking in a crisis and adopted positions derived from the defense of seemingly immutable "principles" such as sovereignty and territorial integrity.

Second, during the Mao and Deng eras, China often displayed a low threshold for the use of limited amounts of force, sometimes seemingly regardless of the human or economic cost involved, and in some cases against, a clearly superior foe. This feature apparently derived primarily from a high level of confidence in the ability to control escalation and a strong belief that a limited application of force was necessary to avoid a larger conflict or to defend core "principles."[31] It also perhaps derived from the ability and willingness of strong leaders such as Mao and Deng to take significant risks.

Third, on balance, China's approach to political-military crises generally has preferred decisive, often coercive (albeit limited) actions over prolonged diplomatic signaling and negotiation or incremental increases in pressure, tit-for-tat moves, etc. Again, this was particularly the case during the Mao and Deng eras.[32] Moreover, crisis signaling generally was designed to communicate resolve, and not to "manage" a crisis.

Fourth, in recent years, the growing influence of public pressure, combined with the likely weakened capacity of senior leaders to resist such pressure, arguably has limited flexibility and self-restraint and distorted signaling in a crisis. This increasingly has become evident during the post-Deng era.

Fifth, an increasingly complex, and fragmented decisionmaking process and a stove-piped intelligence structure have acted to slow down reaction time and also to distort both the accurate assessment of information and clear signaling in a crisis. This also is primarily a feature of the post-Deng era.

On balance, China's approach to crisis management probably has most closely approximated the Type B bargaining "code" or cognitive prism summarized above. As suggested, this code is not regarded as highly conducive to "successful" crisis management practices. Moreover, several of the other largely structural features of China's crisis management behavior—such as aspects of the decisionmaking process and the use of intelligence and information—have served to reinforce such "dysfunctional" features.

However, the question arises as to whether the post-Deng (and now the post-Jiang Zemin) era will result in a significant alteration of China's past approach to crisis management due to both internal and external changes. As suggested by the above discussion, prior to the mid-1990s, most political-military crises were influenced decisively by the existence of:

- A strong, dominant leader;
- Leadership confidence in the ability to prevail militarily in most border areas and in the ability to control escalation;
- Strong controls over public attitudes and behavior; and,
- A low level of dependence on external powers for economic growth and stability.

As indicated, all of these areas have experienced marked change in recent years.

Moreover, before the mid-1990s, most political-military crises involved:

- Limited, largely political objectives for relatively low stakes or in one instance (the Korean War) a clear, unambiguous threat to agreed-upon vital national interests requiring a vigorous response;
- Relatively clear, limited rules of engagement; and,
- Little evidence of external time pressures at the outset of crises.

However, in the future, political-military crises—and a crisis over Taiwan in particular—will likely present significantly different (and in some instances entirely unprecedented) conditions and some major uncertainties not evident in the past.

SECTION 3. "MANAGING" A FUTURE TAIWAN CRISIS. A DUBIOUS PROPOSITION?

In China today, the absence of a charismatic, clearly dominant leader argues in favor of significant levels of caution toward precipitating a crisis and, once started, toward escalating a crisis. As indicated above, unlike Mao and Deng, China's current leadership has less ability to survive major policy errors and hence would presumably treat any crisis over Taiwan with significant caution. Such caution would be reinforced by the huge economic and social damage that could result from a perceived failure to manage a Taiwan crisis, given China's extensive and deep involvement in the global economic order and its heavy reliance on the U.S. trade and investment markets for the maintenance of the high growth regarded as essential to China's future stability.

Second, for China, high barriers likely exist to the success of many deterrence and compellence strategies regarding Taiwan involving the threat or use of limited force. China would find it extremely difficult to attain "local superiority" in a Taiwan crisis, due to both the geography of the area and the nature of the adversary. China's tactical and strategic assets are likely to be highly vulnerable to U.S. conventional stand-off weapons. Moreover, the barrier presented by the Taiwan Strait, combined with U.S. command, control, communications, computer, intelligence, and strategic reconnaissance (C⁴ISR) assets, would make it extremely difficult for China to achieve deception and denial and to act decisively to gain the initiative. And it is difficult for China to assess the likelihood of success against U.S. forces of its key weapons (e.g., ballistic missiles or information warfare), all of which remain largely untested in combat. All of these factors argue in favor of considerable caution toward both precipitating a crisis and in the choice of strategies to manage a crisis.

On the negative side, given the very high stakes involved for China (involving questions of territorial sovereignty, regime legitimacy, and hence social order) and the increased influence of public opinion, the senior Chinese leadership probably would feel enormous pressure to communicate resolve very strongly in a Taiwan crisis and to resist any actions that might suggest capitulation to foreign (read U.S.) pressure. This inclination could increase the chances that China would fall into a classic commitment trap, in which Beijing is compelled to utter ultimatums and then to act on them if deterrence fails.[33] As a result of this attitude, China might find it very difficult to make trade-offs as part of an effort to manage an emerging crisis over Taiwan. For example, it would likely find it extremely hard to adopt an incremental, tit-for-tat approach in a crisis or to trade closer cross-Strait political contacts for even a *perceived* loss of sovereignty claims over Taiwan. Moreover, as indicated above, China has rarely employed such an incrementalist approach in the past.

Such rigidity and the resulting dangers it poses for crisis management could be accentuated by China's continued belief that, in the final analysis, Washington is more likely to "back down" in a Taiwan crisis than Beijing, as mentioned above. Moreover, the above-mentioned disadvantages of achieving local superiority are at least partly counterbalanced by the Chinese view that the United States is vulnerable in high-tech areas and limited by a fear of casualties and the prospect of a prolonged conflict. All of these factors could lead China to miscalculate that it might prevail in a crisis over Taiwan if it can communicate its allegedly stronger resolve clearly and credibly. And China might initiate major military action, even with high risks, if its leaders believe that a closing window of opportunity exists to control or resolve the Taiwan situation.

Third, once in a crisis, China (and the United States) might have great difficulty controlling escalation. In particular, a crisis could move quickly into a conflict that threatens to explode into a larger war. This would certainly be the case if China were to fall into the commitment trap discussed above. China's strong need to communicate resolve, the few nonmilitary means of doing so, and a belief in some key U.S. vulnerabilities could all result in the use of relatively high-risk bargaining strategies by Beijing. It would almost certainly prove difficult for Beijing to convey resolve in any compellance situation regarding Taiwan. Beijing might calculate that the best compellance bargaining strategy would be the *fait accompli*, requiring a rapid military strike that could increase greatly the chances of an all-out conflict.

In addition, a Taiwan crisis probably would offer less time for assessment and negotiation, and would increase the chance that military actions of various types would short-circuit diplomatic or political options, thus affecting efforts at escalation control. This danger could be accentuated, in the Chinese case, by the apparent existence of weak horizontal lines of communication within the People's Liberation Army (PLA) and between the PLA and the civilian sides of government. It also might be more difficult to read and control signals and moves in a Taiwan crisis. For example, China might view strong U.S. military assistance to Taiwan in the opening days of a crisis as equivalent to a "first shot" escalation requiring a vigorous response. Even more serious, in an escalating crisis, China might interpret apparent U.S. attacks on key Chinese command and control facilities or military assets relevant to the PLA's prosecution of strikes against Taiwan as a threat against Beijing's larger conventional and strategic capabilities and respond accordingly.

In general, it would likely prove difficult to read signals accurately, divide-up issues, make trade-offs, and accept short-term losses in a Taiwan crisis. In addition, the existence of a third party with independent interests and policy options could greatly complicate efforts at crisis management by both the United States and China. Taiwan's behavior could produce significant misperceptions, resulting in unwanted escalation. Finally, China might miscalculate the risks involved in a Taiwan crisis by assuming that it could apply pressure on Taiwan to deter U.S. military intervention.

SECTION 4. QUESTIONS AND CONCLUSIONS

The above analysis of China's crisis management behavior, in the context provided by a larger framework for understanding crisis management, raises several important questions for further examination.[34]

1. What difference did Mao Zedong make to the crisis decisionmaking process? Was Mao more willing to take risks, less "rational," more ideological, and more rigid and resistant to negotiation? Was he more able to endure failure or loss of prestige in a crisis and less susceptible to public pressure? How would an individual with Mao's strong personality and views operate in China's current crisis decisionmaking system?

2. Does China's current image of the United States reduce available options in crisis bargaining, lower self-restraint, and make goals more "absolute" in a crisis, thus making it more difficult to achieve a resolution?

3. How does historical memory matter? Does it explain how elites understand a crisis? Or how the public understands a crisis? Since historical memory is to a great extent created by educational systems and propaganda systems, then it can presumably be changed. What is involved in reducing the impact of China's current historical memory relevant to political-military crises?

4. The United States is viewed by many Chinese as possessing a much larger variety of means to manage a crisis. At the same time, it is presumably constrained by a fear of casualties, prolonged conflict, and economic costs. Moreover, the United States is viewed by China as likely having less at stake in a future conflict with China. For all these reasons, many Chinese apparently believe that the United States, in most instances, can choose more easily to avoid (or back away from) a crisis with China involving the use of force. This suggests, to some Chinese, that the United States is likely to be deterrable more easily in a crisis than China. What are the concrete policy implications of these asymmetries for managing future crises? How do these asymmetries shape how China's leaders assess their options and positions in a crisis? How do they shape approaches to different types of crises, such as "sudden incidents" versus other types of crises? What type of structures or processes can be created to mitigate the effects of these asymmetries?

   Moreover, how does the superior power and leverage of the United States affect this calculus? The United States might believe that its superior power and leverage would allow it to prevail, even if the stakes are not as important for it as they are for China. And how does the need to maintain credibility as a superpower and the need to retain domestic support in a democratic system affect this calculus? U.S. leaders might judge that a crisis could undermine their political position, as well as the U.S. ability to exercise its global responsibilities, and thus be less willing to "back down" in a crisis than China's leaders might think.

5. Which emotions are most common in Chinese decisionmaking: Nationalism? Hatred? Fear of loss of political power? Face? What options are completely ruled out? Are they ruled out because they would bring "hardliners" further into the process?

6. How is the principled status of an adversary (enemy, friend, or neither) determined? What is the link between each status and the way options are assessed in a crisis? From China's

perspective, what crises were avoided in the first place because of the "friendly" state of relations between the United States and China?

7. The first component of the notion of *youli, youli, youjie* outlines the defensive nature of the conditions under which China might use force in a crisis. However, what often counts as a provocation appears to be a matter of principle and not necessarily a type of military action. It seems that China has struck—and may strike—first against what it perceives to be a challenge to its interests, even if it is not attacked militarily. What can be considered a provocation in the absence of a military attack? What other principles are invoked to determine when a provocation occurs? How can perceived violations of these principles be managed during a crisis?

8. In past political-military crises under Mao Zedong and Deng Xiaoping, the leadership showed a clear willingness to sustain significant military and/or economic costs, if it were confident that, by doing so, it could attain its core (usually political) objectives. Do Chinese leaders hold the same general viewpoint toward costs and objectives today? Many Chinese observers insist that China's risk calculus has changed significantly since the Mao and Deng eras, and that the use of force is not regarded as an effective tool to achieve limited political gains in a crisis. However, other Chinese seem to disagree. If the use of the military is regarded by China's present leaders as a last resort (due to the changed nature of the Chinese regime and the emergence of a different risk calculus), how does China signal resolve and create stability, especially regarding a sensitive territorial issue such as Taiwan? How can one stop a relatively small incident (i.e., political moves by Taiwan) from becoming a major threat and a crisis? Will China rely more on nonmilitary levers, or on U.S. (or UN) actions to prevent the deterioration of the Taiwan situation? Or will China still use limited force to convey resolve, which could be misinterpreted as an attempt to "win"?

9. Would the risks involved in employing military alerts or other potentially dangerous means to signal resolve be much lower

in the case of a Sino-U.S. crisis than during the Cold War, when U.S. and Soviet nuclear forces were coupled tightly and wedded to warfighting strategies?

10. Despite an emphasis on clear signaling for deterrence purposes, China has shown some hesitation over the establishment of a direct line of communication with the United States through a hotline. Can a mechanism and procedure be developed that would overcome (or at least appreciably reduce) such concerns and enhance the positive effect of signaling?

11. Some analysts of crisis behavior believe that the United States can signal resolve more easily in a crisis not only because of its superior military capabilities, but also because of the supposedly high cost involved in retreating from a position, once established. This is due to a belief that, because the United States is a democracy, political leaders fear that they will suffer in the next election if they do not stand by their commitments. In other words, because so-called audience costs are higher in a democracy, signals from democratic leaders are believed to be much stronger or potentially more costly if unsuccessful, than signals from nondemocracies. This suggests that it might be harder for the United States to back down in a crisis, which contradicts what appears to be the Chinese assumption of a lower U.S. stake in a Sino-U.S. crisis and hence a greater willingness or ability to compromise in a crisis. Is this argument, and the resulting assertion that democracies have greater credibility in conveying resolve and commitment, accepted by the Chinese?

12. Is China aware of the danger of the commitment trap? Or, putting it another way, does China's past tendency to regard crises in zero-sum terms, its resistance to appearing weak in a crisis (given public and elite opinion), and its belief that many crises involve very high stakes together reduce its concerns over falling into such a trap?

13. Different U.S. and Chinese assessments of the purpose, motivations, and influence of each side's signals easily can lead to misinterpretations. How can each side send signals

that convey resolve (or accommodation) without producing an overreaction, leading to unwanted escalation (or, in the case of a signal of accommodation, an inaccurate assumption of weakness by the adversary)? What do the Chinese and U.S. sides pay attention to in order to determine the other side's will and intentions?

14. Specifically, what signals should be considered as indications of a willingness to cooperate or conciliate, and which should be considered conflictual? How can China and the United States credibly signal restraint in the use of force? Are there strong incentives not to communicate clear or honest signals, or to delay the sending of signals?

15. How have contact and communication among and between national security elites in both China and the United States affected the duration and severity of crises? Is contact necessary?

16. Some analysts believe that political-military crises have been used by past Chinese leaders to build support for the government among the Chinese populace and within the elite. If the "safe" position for Chinese leaders remains the hard-line position, and the Chinese regime needs to strengthen its legitimacy by appealing to nationalist sentiments, then is the temptation to use a crisis to build support as great as or greater than arguably existed during the Maoist era?

17. Why is the Chinese government more sensitive to public opinion than in the past? Specifically, how does public opinion influence crisis decisionmaking? Can its effects be limited or controlled? Does the Chinese government actually understand public opinion? Have Chinese leaders overestimated the degree to which the Chinese public cares about Taiwan independence, for example?

18. There is little doubt that the role and influence of military plans in Chinese policymaking should be examined more closely to understand their role in a crisis. How do existing or predetermined PLA plans influence policymaking and crisis management? How does the PLA understand high-level

political decisionmaking in a crisis? Are civilians aware of the diplomatic implications of military plans? Do military plans reduce or lengthen the time pressure in the crisis?

19. As noted above, special task forces have reportedly been formed during recent crises. Are these important (and new?) mechanisms designed to improve crisis decisionmaking?

20. What role, if any, has international law played over time as a constraint regarding Chinese calculations of the use of force and the management of crises? As China becomes more integrated in the international economy and sees itself as a responsible great power, has international law increasingly become a constraint in the handling of crises?

Conclusions.

From the above preliminary assessment of scholarly studies of crisis management and China's past crisis management behavior, and from the many key questions that remain regarding Chinese crisis behavior, one can draw the following overall conclusions.

First, China's perceptions and behavior during past political-military crises—and especially those crises that occurred during the Mao and Deng eras—provide a very limited foundation for our understanding of Beijing's future approach to crisis management. In many respects, most of the key variables influencing Chinese behavior are undergoing profound change and apparently new variables (such as the influence of independent public sentiment) have emerged. Thus, it is essential to determine which features endure and which are new or evolving, and in what manner. Although this chapter provides some hints at an answer in some of these areas, much more work needs to be done.

Second, what we know about China's past approach to crises, as well as some of the features of China's current crisis decisionmaking system, are not terribly reassuring to advocates of "prudent" crisis management behavior, particularly when one considers the special— and in many ways unprecedented—context of a possible future crisis over Taiwan. The danger exists that China's leaders might miscalculate in a variety of significant ways both before and during such a crisis. On the other hand, considerable constraints apparently exist against high risk behavior by the Chinese leadership, given the stakes and the uncertainties involved in a Taiwan crisis. Scholars should attempt to more accurately assess the relative balance that exists between these negative and positive aspects of Chinese crisis behavior regarding Taiwan.

Third, as indicated by the above 20 questions, many, if not most, of the key features relevant to the above five categories of variables that influence China's crisis management behavior remain unknown to outside observers, and even to many Chinese observers. Most of the above questions can be translated into research designs and explored systematically, in some cases through the use of surveys. This would be a major task, but one of great relevance not only to existing scholarship on crisis management but, more important, to the handling of critical policy challenges that could influence the prospects for war and peace between China and the United States.

Fourth, some elements of China's approach to crises and crisis management ultimately are unknowable, especially prior to a crisis, and even to the Chinese themselves. As the scholarly literature on crisis management indicates, actual crisis behavior is very context-dependent and will vary according to many factors, including the specific leaders involved in a crisis, the exact nature (e.g., origin and scope) of a crisis, and, of course, the behavior of outside actors and events. Moreover, in general, it is virtually impossible to know, both before and during a crisis, critical issues such as precisely what every signal means, what level of resolve exists within the Chinese leadership at any given time, and how domestic versus external factors are influencing behavior. At best, as the above examination of crisis bargaining codes suggests, we can expand our understanding of certain broad tendencies that influence China's crisis management behavior, thereby reducing the likelihood of undesirable outcomes—such as inadvertent war—for leaders on all sides. This alone would be a significant achievement.

APPENDIX I

MAJOR SOURCES ON POLITICAL-MILITARY CRISES AND CRISIS MANAGEMENT

Jacob Bercovitch, "The Structure and Diversity of Mediation in International Relations," in Jacob Bercovitch and Jeffrey Z. Rubin, eds., *Mediation in International Relations*, New York: St. Martin's Press, 1992.

Jacob Bercovitch and Patrick M. Regan, "Managing Risks in International Relations: The Mediation of Enduring Rivalries," in Gerald Schneider and Patricia A. Weitsman, eds., *Enforcing Cooperation*, New York: St. Martin's Press, 1997.

Michael Brecher and Jonathan Wilkenfeld, *Crisis, Conflict, and Instability*, Oxford: Pergamon, 1989.

Michael Brecher and Jonathan Wilkenfeld, *A Study of Crisis*, Ann Arbor: The University of Michigan Press, 2000.

Michael Brecher, Jonathan Wilkenfeld, and Sheila Moser, *Crises in the Twentieth Century: Handbook of International Crises*, Oxford: Pergamon, 1988.

Bruce Bueno de Mesquita, *The War Trap*, New Haven: Yale University Press, 1981.

Bruce Bueno de Mesquita and David Lalman, *War and Reason*, New Haven: Yale University Press, 1992.

John W. Burton, *Conflict: Resolution and Prevention*, New York: St. Martin's Press, 1990.

George W. Downs, "The Rational Deterrence Debate," *World Politics*, Vol. 41, No. 2, January 1989.

John Garnett, *Theories of Peace and Security*, London: Macmillan, 1970.

Alexander L. George, eds., *Avoiding War: Problems of Crisis Management*, Boulder, CO: Westview Press, 1991.

Alexander L. George, "Crisis Management: The Interaction of Political and Military Considerations," *Survival*, Vol. 26, No. 5, September/October 1984.

Alexander L. George and David K. Simons, *The Limits of Coercive Diplomacy*, 2nd ed., Boulder, CO: Westview Press, 1994.

Alexander L. George and Richard Smoke, *Deterrence in American Foreign Policy*, New York: Columbia University Press, 1974.

Arthur N. Gilbert and Paul Gordon Lauren, "Crisis Management: An Assessment and Critique," *Journal of Conflict Resolution*, Vol. 24, No. 4, December 1980.

Ernest B. Haas, "Regime Decay: Conflict Management and International Organizations, 1945-1981," *International Organization*, Vol. 37, No. 2, April 1983.

Ole R. Holsti, "Crisis Decisionmaking," in P. E. Tetlock, *et al.*, eds., *Behavior, Society and Nuclear War*, New York: Oxford University Press, 1989.

Ole R. Holsti, *Crisis, Escalation, War*, Montreal: McGill-Queen's University Press, 1972.

Ole R. Holsti, "Crisis Management," in Betty Glad, ed., *Psychological Dimensions of War*, Newbury Park, CA: Sage, 1990.

Paul Huth, Christopher Gelpi, and D. Scott Bennett, "The Escalation of Great Power Militarized Disputes: Testing Rational Deterrence Theory and Structural Realism," *American Political Science Review*, Vol. 87, No. 3, September 1993.

Paul Huth and Bruce Russett, "Deterrence Failure and Crisis Escalation," *International Studies Quarterly*, Vol. 32, No. 1, March 1988.

Paul Huth and Bruce Russett, "Testing Deterrence Theory: Rigor Makes a Difference," *World Politics*, Vol. 42, No. 4, July 1990.

Robert Jervis, *Perception and Misperception in International Politics*, Princeton: Princeton University Press, 1976.

Robert Jervis, "Rational Deterrence: Theory and Evidence," *World Politics*, Vol. 41, No. 2, January 1989.

Richard Ned Lebow, *Between Peace and War: The Nature of International Crisis*, Baltimore: The Johns Hopkins University Press, 1981.

Richard Ned Lebow, *Nuclear Crisis Management*, Ithaca, NY: Cornell University Press, 1987.

Richard Ned Lebow and Janice Gross Stein, "Beyond Deterrence," *Journal of Social Issues*, Vol. 43, No. 4, Winter 1987.

Richard Ned Lebow and Janice Gross Stein, "Deterrence: The Elusive Dependent Variable," *World Politics*, Vol. 42, No. 3, April 1990.

Russell J. Leng, "Influence Techniques among Nations," in Philip E. Tetlock, *et al.*, eds., *Behavior, Society, and International Conflict*, New York: Oxford University Press, 1993.

Russell J. Leng, "When Will They Ever Learn: Coercive Bargaining in Recurrent Crises," *Journal of Conflict Resolution*, Vol. 27, No. 3, September 1983.

Charles Lockhart, *Bargaining in International Crises*, New York: Columbia University Press, 1979.

Zeev Maoz, "Crisis Initiation: A Theoretical Exploration of a Neglected Topic in International Crisis Theory," *Review of International Studies*, Vol. 8, No. 4, October 1982.

Charles A. McClelland, "The Acute International Crisis," in Klaus Knorr and Sidney Verba, eds., *The International System: Theoretical Essays*, Princeton: Princeton University Press, 1961.

Charles A. McClelland, "The Beginning, Duration, and Abatement of International Crises: Comparisons in Two Conflict Arenas," in C.F. Hermann, ed., *International Crises: Insights and Evidence*, New York: The Free Press, 1972.

John Mearsheimer, *Conventional Deterrence*, Ithaca, NY: Cornell University Press, 1983.

Benjamin Miller, "Explaining Great Power Cooperation in Conflict Management," *World Politics*, Vol. 45, No. 1, October 1992.

Patrick Morgan, *Deterrence: A Conceptual Analysis*, Beverly Hills: Sage, 1977.

T. Clifton Morgan, *Untying the Knot of War: A Bargaining Theory of International Crises*, Ann Arbor: The University of Michigan Press, 1994.

James D. Morrow, "Signaling Difficulties with Linkage in Crisis Bargaining," *International Studies Quarterly*, Vol. 36, No. 1, March 1992.

Barry O'Neill, "Game Theory and the Study of the Deterrence of War," in Paul C. Stern, *et al.*, eds., *Perspectives on Deterrence*, New York: Oxford University Press, 1989.

J. Philip Rogers, "Crisis Bargaining Codes and Crisis Management," in Alexander L. George, ed., *Avoiding War: Problems of Crisis Management*, Boulder, CO: Westview Press, 1991.

Alvin Roth, ed., *Game-Theoretic Models of Bargaining*, New York: Cambridge University Press, 1985.

Bruce Russett, "The Calculus of Deterrence," *Journal of Conflict Resolution*, Vol. 7, No. 2, June 1963.

Thomas C. Schelling, *The Strategy of Conflict*, Cambridge: Harvard University Press, 1960.

Glenn H. Snyder and Paul Diesing, *Conflict Among Nations: Bargaining, Decisionmaking, and System Structure in International Crises*, Princeton: Princeton University Press, 1977.

Glenn H. Snyder, *Deterrence and Defense*, Princeton: Princeton University Press, 1961.

Richard C. Snyder, H. W. Bruck, and Burton Sapin, "Decisionmaking as an Approach to the Study of International Politics," in Richard C. Snyder, H. W. Bruck, and Burton Sapin, eds., *Foreign Policy Decisionmaking: An Approach to the Study of International Politics*, New York: The Free Press, 1962.

John Steinbrunner, "Beyond Rational Deterrence," *World Politics*, Vol. 28, No. 4, July 1976.

Peter F. Trumbore and Mark A. Boyer, "Two-Level Negotiations in International Crisis: Testing the Impact of Regime Type and Issue Area," *Journal of Peace Research*, Vol. 37, No. 6, November 2000.

R. Harrison Wagner, "Deterrence and Bargaining," *Journal of Conflict Resolution*, Vol. 26, No. 2, June 1982.

R. Harrison Wagner, "Rationality and Misperception in Deterrence Theory," *Journal of Theoretical Politics*, Vol. 4, No. 2, April 1992.

R. Harrison Wagner, "Uncertainty, Rational Learning, and Bargaining in the Cuban Missile Crisis," in Peter C. Ordeshook, ed., *Models of Strategic Choice in Politics*, Ann Arbor: The University of Michigan Press, 1989.

Jonathan Wilkenfeld, "Trigger-Response Transitions in Foreign Policy Crises, 1929-1985," *Journal of Conflict Resolution*, Vol. 35, No. 1, March 1991.

Jonathan Wilkenfeld, Michael Brecher, and Sheila Moser, *Crises in the Twentieth Century: Handbook of Foreign Policy Crises*, Oxford: Pergamon, 1988.

Jonathan Wilkenfeld, Kathleen Young, Victor Asal, and David Quinn, "Mediating International Crises: Cross-National and Experimental Perspectives," *Journal of Conflict Resolution*, Vol. 47, No. 3, September 2003.

Phil Williams, *Crisis Management: Confrontation and Diplomacy in the Nuclear Age*, London: Martin Robertson, 1976.

Samuel S. G. Wu, "To Attack or Not to Attack," *Journal of Conflict Resolution*, Vol. 34, No. 3, September 1990.

H. Peyton Young, ed., *Negotiation Analysis*, Ann Arbor: The University of Michigan Press, 1991.

Frank C. Zagare, "Classical Deterrence Theory: A Critical Assessment," *International Interactions*, Vol. 21, No. 4, October/December 1996.

Frank C. Zagare, "Rationality and Deterrence," *World Politics*, Vol. 42, No. 2, January 1990.

I. William Zartman, "Alternative Attempts at Crisis Management: Concepts and Processes," in Gilbert R. Winham, ed., *New Issues in International Crisis Management*, Boulder, CO: Westview Press, 1988.

APPENDIX II

MAJOR SOURCES ON CHINESE CONFLICT BEHAVIOR, INTERSTATE BARGAINING, CRISIS MANAGEMENT, AND STRATEGIC VIEWS AND BELIEFS

Jonathan Adelman and Shih Chih-yu, *Symbolic War: The Chinese Use of Force, 1840-1980*, Taipei: National Chengchi University, 1993.

David Bachman, "Domestic Sources of Chinese Foreign Policy," in Samuel Kim, ed., *China and the World: Chinese Foreign Relations in the Post-Cold War Era*, Boulder, CO: Westview Press, 1994.

Davis B. Bobrow, "The Chinese Communist Conflict System," *Orbis*, Vol. 9, No. 1, Winter 1966.

Davis B. Bobrow, "Peking's Military Calculus," in Davis B. Bobrow, ed., *Components of Defense Policy*, Chicago: Rand McNally, 1965.

Davis B. Bobrow, Steve Chan, and John A. Kringen, *Understanding Foreign Policy Decisions: The Chinese Case*, New York: Free Press, 1979.

Mark Burles and Abram N. Shulsky, *Patterns in China's Use of Force: Evidence from History and Doctrinal Writings*, Santa Monica: RAND, 2000.

Steve Chan, "Chinese Conflict Calculus and Behavior: Assessment from a Perspective of Conflict Management," *World Politics*, Vol. 30, No. 3, April 1978.

Chen Bingde, "Intensify Study of Military Theory to Ensure Quality of Army Building: Learning from Thought and Practice of the Core of the Three Generations of Party Leadership in Studying Military Theory," in *Zhongguo Junshi Kexue*, No. 3, August 20, 1997, in *Foreign Broadcast Information Service—China* (henceforth *FBIS-CHI*), March 6, 1998.

Chen Jian, *China's Road to the Korean War: The Making of the Sino-American Confrontation*, New York: Columbia University Press, 1994.

King C. Chen, *China's War With Vietnam, 1979: Issues, Decisions, and Implications*, Stanford: Hoover Institution Press, 1987.

Thomas J. Christensen, "China, the U.S.-Japan Alliance and the Security Dilemma in East Asia," *International Security*, Vol. 23, No. 4, Spring 1999.

Thomas J. Christensen, "Chinese Realpolitik," *Foreign Affairs*, Vol. 75, No. 5, September/October 1996.

Thomas J. Christensen, *Useful Adversaries: Grand Strategy, Domestic Mobilization, and Sino-American Conflict, 1947-1958*, Princeton: Princeton University Press, 1996.

Arthur Cohen, "The Sino-Soviet Border Crisis," in Alexander L. George, ed., *Avoiding War: Problems of Crisis Management*, Boulder, CO: Westview Press, 1991.

David M. Finkelstein, "China's National Military Strategy," in James C. Mulvenon and Richard H. Yang, eds., *The People's Liberation Army in the Information Age*, Santa Monica: RAND, 1999.

Paul H. B. Godwin, "Changing Concepts of Doctrine, Strategy and Operations in the Chinese People's Liberation Army 1978–87," *China Quarterly*, No. 112, December 1987.

Paul H.B. Godwin, "Chinese Military Strategy Revised: Local and Limited War," *The Annals of the American Academy of Political Science*, Vol. 519, January 1992.

Paul H.B. Godwin, "Force Projection and China's National Military Strategy," in C. Dennison Lane, Mark Weisenbloom, and Simon Li, eds., *Chinese Military Modernization*, London: Kegan Paul, 1996.

Paul H.B. Godwin, "From Continent to Periphery: PLA Doctrine, Strategy and Capabilities Towards 2000," *China Quarterly*, No. 146, June 1996.

Melvin Gurtov and Byong-Moo Hwang, *China Under Threat: The Politics of Strategy and Diplomacy*, Baltimore: The Johns Hopkins University Press, 1980.

Carol Lee Hamrin and Zhao Suisheng, *Decisionmaking in Deng's China: Perspectives From Insiders*, Armonk, NY: M. E. Sharpe, 1995.

Hao Yufan and Zhao Zhihai, "China's Decision to Enter the Korean War," *China Quarterly*, No. 121, March 1990.

He Di, "The Evolution of the People's Republic of China's Policy Toward the Offshore Islands," in Warren I. Cohen and Akira Iriye, eds., *The Great Powers in East Asia, 1953-1960*, New York: Columbia University Press, 1990.

Huang Jialun, "Three-Point Thinking on Developing Combat Theory," *Jiefangjun Bao*, April 7, 1998, in *FBIS-CHI*, May 1, 1998.

Huang Xing and Zuo Quandian, "Holding the Initiative in Our Hands in Conducting Operations, Giving Full Play to Our Own Advantages to Defeat Our Enemy: A Study of the Core Idea of the Operational Doctrine of the People's Liberation Army," *Zhongguo Junshi Kexue*, No. 4, November 20, 1996, in *FBIS-CHI*, June 17, 1997.

Jia Wenxian, Zheng Shouqi, and Guo Weimin, "Tentative Discussion of Special Principles of a Future Chinese Limited War," *Guofang Daxue Xuebao*, No. 11, November 1, 1987, in *Joint Publications Research Service* (henceforth *JPRS*), July 12, 1988.

Jiao Wu and Xiao Hui, "Modern Limited War Calls for Reform of Traditional Military Principles," *Guofang Daxue Xuebao*, No. 11, November 1, 1987, in *JPRS*, July 12, 1988.

Ellis Joffe, "People's War Under Modern Conditions: A Doctrine for Modern War," *China Quarterly*, No. 112, December 1987.

Alastair Iain Johnston, "China's Militarized Interstate Dispute Behavior 1949-1992: A First Cut at the Data," *China Quarterly*, No. 153, March 1998.

Alastair Iain Johnston, "China's New 'Old Thinking': The Concept of Limited Deterrence," *International Security*, Vol. 20, No. 3, Winter 1995/96.

Alastair Iain Johnston, *Cultural Realism: Strategic Culture and Grand Strategy in Chinese History*, Princeton: Princeton University Press, 1995.

Arthur Lall, *How Communist China Negotiates*, New York: Columbia University Press, 1968.

David M. Lampton, *The Making of Chinese Foreign and Security Policy in the Era of Reform: 1978-2000*, Stanford: Stanford University Press, 2001.

Nathan Leites, *The Operational Code of the Politburo*, New York: McGraw-Hill Book Co., 1951.

Li Nan, "The PLA's Warfighting Doctrine, Strategy, and Tactics," *China Quarterly*, No. 146, June 1996.

Li Yaqiang, "Will Large-Scale Naval Warfare Recur?" *Jianchuan Zhishi*, August 8, 1995, in *FBIS-CHI*, August 8, 1996.

Lin Chong-pin, *China's Nuclear Weapons Strategy: Tradition Within Evolution*, Lexington, MA: D. C. Heath, 1988.

Liu Fengcheng, "Concentrate Forces in New Ways in Modern Warfare," *Jiefangjun Bao*, November 21, 1995, in *FBIS-CHI*, January 29, 1996.

Liu Huaqing, "Unswervingly Advance Along the Road of Building a Modern Army with Chinese Characteristics," *Jiefangjun Bao*, August 6, 1993, in *FBIS-CHI*, August 18, 1993.

Lu Linzhi, "Preemptive Strikes Crucial in Limited High-Tech Wars," *Jiefangjun Bao*, February 14, 1996, in *FBIS-CHI*, February 14, 1996.

Lu Ning, *The Dynamics of Foreign-Policy Decisionmaking in China*, Boulder, CO: Westview Press, 1997.

Neville Maxwell, *India's China War*, New York: Random House, 1970.

James C. Mulvenon, "The Limits of Coercive Diplomacy: The 1979 Sino-Vietnamese Border War," *Journal of Northeast Asian Studies*, Vol. 14, No. 3, Fall 1995.

Andrew J. Nathan and Robert S. Ross, *The Great Wall and the Empty Fortress: China's Search for Security*, New York: Norton Press, 1997.

Glenn D. Paige, *The Korean Decision*, New York: The Free Press, 1968.

Pan Shiying, "Thoughts About the Principal Contradiction in Our Country's National Defense Construction," *Jiefangjun Bao*, April 14, 1987, in *FBIS-CHI*, April 29, 1987.

Michael Pillsbury, *Dangerous Chinese Misperceptions: Implications for DoD*, Washington, DC: Office of Net Assessment, 1998.

Jonathan D. Pollack, *Security, Strategy, and the Logic of Chinese Foreign Policy*, Berkeley: University of California at Berkeley, Institute of East Asian Studies, 1981.

Ren Yue, "China's Dilemma in Cross-Strait Crisis Management," *Asian Affairs*, Vol. 24, No. 3, Fall 1997.

Thomas W. Robinson, "The Sino-Soviet Border Dispute: Background, Development, and the March 1969 Clashes," *American Political Science Review*, Vol. 66, No. 4, December 1972.

Robert S. Ross, *The Indochina Tangle: China's Vietnam Policy, 1975–1979*, New York: Columbia University Press, 1988.

Marl Ryan, *Chinese Attitudes Toward Nuclear Weapons*, Armonk, NY: M. E. Sharpe, 1989.

Louis J. Samuelson, *Soviet and Chinese Negotiating Behavior: The Western View*, Beverly Hills: Sage, 1976.

Andrew Scobell, *China and Strategic Culture*, Carlisle Barracks, PA: Strategic Studies Institute, U.S. Army War College, 2002.

Andrew Scobell, *China's Use of Military Force: Beyond the Great Wall and the Long March*, Cambridge and New York: Cambridge University Press, 2003.

Gerald Segal, *Defending China*, Oxford: Oxford University Press, 1985.

Shi Yukun, "Lt. Gen. Li Jijun Answers Questions on Nuclear Deterrence, Nation-state, and Information Age," *Zhongguo Junshi Kexue*, No. 3, August 20, 1995, in *FBIS-CHI*, August 20, 1995.

Abram N. Shulsky, *Deterrence Theory and Chinese Behavior*, Santa Monica: RAND, 2000.

Richard Solomon, *Chinese Negotiating Behavior: Pursuing Interests Through "Old Friends,"* Washington, DC: United States Institute of Peace Press, 1999.

Sun Zi'an, "Strategies to Minimize High-Tech Edge of Enemy," *Xiandai Bingqi*, No. 8, August 8, 1995, in *FBIS-CHI*, March 6, 1996.

Michael D. Swaine and Ashley J. Tellis, *Interpreting China's Grand Strategy: Past, Present, and Future*, Santa Monica: RAND, 2000.

Tang Guanghui, "An Analysis of Post–Cold War Security," *Shijie Zhishi*, No. 19, October 1, 1996.

Tang Tsou and Morton Halperin, "Mao Tse-tung's Revolutionary Strategy and Peking's International Behavior," *American Political Science Review*, Vol. 59, No. 1, March 1965.

Wang Chunyin, "Characteristics of Strategic Initiative in High-Tech Local Wars," *Hsien-Tai Chun-Shih*, No. 245, June 11, 1997, in *FBIS-CHI*, October 30, 1997.

Allen S. Whiting, *China Crosses the Yalu: The Decision to Enter the Korean War*, New York: The Macmillan Company, 1960.

Allen S. Whiting, *The Chinese Calculus of Deterrence: India and Indochina*, Ann Arbor: The University of Michigan Press, 1975.

Allen S. Whiting, "The Use of Force in Foreign Policy by the People's Republic of China," *The Annals of the American Academy of Political and Social Science*, Vol. 402, July 1972.

Xia Liping, "Theory and Practice of Crisis Management in the United States—Sino-US Relations as an Example," *Beijing Meiguo Yanjiu*, Vol. 17, No. 2, June 2003.

Yan Youqiang and Chen Rongxing, "On Maritime Strategy and the Marine Environment," *Zhongguo Junshi Kexue*, No. 2, May 20, 1997, in *FBIS-CHI*, October 14, 1997.

Yang Dezhi, "A Strategic Decision on Strengthening the Building of Our Army in the New Period," *Hongqi*, No. 15, August 1, 1985, in *FBIS-CHI*, August 8, 1985.

Yu Guohua, "On Turning Strong Force Into Weak and Vice-Versa in a High-Tech Local War," *Zhongguo Junshi Kexue*, No. 2, May 20, 1996, in *FBIS-CHI*, May 20, 1996.

Zhang Shu Guang, *Deterrence and Strategic Culture: Chinese American Confrontations, 1949-1958*, Ithaca, NY: Cornell University Press, 1992.

Zhang Shu Guang, *Mao's Military Romanticism: China and the Korean War, 1950-1953*, Lawrence: University Press of Kansas, 1995.

Zheng Shenxia and Zhang Changzhi, "Air Power as Centerpiece of Modern Strategy," *Zhongguo Junshi Kexue*, No. 2, May 20, 1996, in *FBIS-CHI*, May 20, 1996.

Zong He, "Tentative Discussion on the Characteristics of Modern Warfare," *Shijie Zhishi*, No. 15, August 1, 1983, in *JPRS*, October 11, 1983.

1. Jonathan Wilkenfeld, "Concepts and Methods in the Study of International Crises," unpublished paper, University of Maryland, February 2004, p. 4. Also see Michael Brecher and Jonathan Wilkenfeld, *A Study of Crisis*, Ann Arbor: The University of Michigan Press, 2000. For a slightly different definition that stresses the element of surprise in the emergence of a crisis, see Charles F. Hermann, "International Crisis as a Situational Variable," in J. N. Rosenau, ed., *International Politics and Foreign Policy*, New York: The Free Press, 1969, p. 414.

2. See Wilkenfeld, "Concepts and Methods," pp. 4-6. The most recent Sino-American political-military crises thus do not all qualify as full crises. For example, only the 1995-96 Taiwan Strait crisis can be considered a full-blown crisis, not the April 2001 EP3 incident or the 1999 Belgrade embassy bombing.

3. Alexander L. George, "A Provisional Theory of Crisis Management," in Alexander L. George, ed., *Avoiding War: Problems of Crisis Management*, Boulder, CO: Westview Press, 1991, pp. 22-23.

4. Glenn H. Snyder and Paul Diesing, *Conflict Among Nations: Bargaining, Decisionmaking, and System Structure in International Crises*, Princeton: Princeton University Press, 1977.

5. As Wilkenfeld points out, jointly opposed situations offer the greatest opportunity of cooperative strategies, while diametrically opposed situations are the most dangerous because they would appear to offer little opportunity for management or resolution. Indeed, diametrically opposed situations are largely zero sum, while quasi-opposed situations are mixed sum, allowing for the possibility of win-win strategies.

6. Crisis management is most often thought of as "the attempt to control events during a crisis to prevent significant and systematic violence from occurring." See Graham Evans and Jeffrey Newnham, *The Penguin Dictionary of International Relations*, London: Penguin Books, 1998, as cited in Wilkenfeld, "Concepts and Methods." Also see John W. Burton, *Resolving Deep-Rooted Conflict: A Handbook*, Lanham, MD: University Press of America, 1987; John W. Burton, *Conflict: Resolution and Prevention*, New York: St. Martin's Press, 1990; Herbert C. Kelman, "Informal Mediation by the Scholar/Practitioner," in Jacob Bercovitch and Jeffrey Z. Rubin, eds., *Mediation in International Relations*, New York: St. Martin's Press, 1992; Marieke Kleiboer, *The Multiple Realities of International Mediation*, Boulder, CO: Lynne Rienner Publishers, 1998; Marieke Klieboer, "Great Power Mediation: Using Leverage to Make Peace?" in Jacob Bercovitch, ed., *Studies in International Mediation: Essays in Honor of Jeffrey Z. Rubin*, Houndsmills, Basingstoke, Hampshire, New York: Palgrave MacMillan, 2002, pp. 127-40; Deborah M. Kolb and Eileen F. Babbitt, "Mediation Practice on the Home Front: Implications for Global Conflict Resolution," in John A. Vasquez, *et al.*, eds., *Beyond Confrontation: Learning Conflict Resolution in the Post-Cold War Era*, Ann Arbor: The University of Michigan Press, 1995, pp. 63-85; Kumar Rupesinghe, "Mediation in Internal Conflicts: Lessons from Sri Lanka," in Jacob Bercovitch, ed., *Resolving International Conflicts: The Theory*

and Practice of Mediation, Boulder, CO: Lynne Rienner Publishers, 1996, pp. 153-69; and I. William Zartman, "Toward the Resolution of International Conflicts," in I. William Zartman and J. Lewis Rasmussen, eds., *Peacemaking in International Conflict: Methods and Techniques*, Washington, DC: United States Institute of Peace Press, 1997, p.11.

7. George, "A Provisional Theory of Crisis Management," pp. 23, 26. See Appendix 1 for other major sources on political-military crises and crisis management.

8. Once an all-out war erupts, crisis management ceases, and efforts at achieving military victory and/or ending hostilities begin. These processes are markedly different in nature from crisis management.

9. As defined in some detail by Alexander George, offensive crisis management strategies include: 1) blackmail; 2) limited, reversible probes; 3) controlled pressure; 4) attempts to establish a fait accompli; and 5) slow attrition. Defensive crisis management strategies include: 1) coercive diplomacy; 2) limited escalation combined with deterrence of counterescalation; 3) tit-for-tat reprisals combined with deterrence of escalation; 4) accepting a "test of capabilities" on conditions set by the challenger(s), to reverse the expected outcome; 5) efforts to draw the line; 6) conveying commitment and resolve to avoid miscalculation; and 7) various time-buying actions and proposals in order to explore a negotiated settlement. George, "Strategies for Crisis Management," in George, ed., *Avoiding War: Problems of Crisis Management*, pp. 379-393.

10. Alexander L. George, "Findings and Recommendations," in George, ed., *Avoiding War: Problems of Crisis Management*, pp. 557-558.

11. See J. Philip Rogers, "Crisis Bargaining Codes and Crisis Management," in George, ed., *Avoiding War: Problems of Crisis Management*, Boulder, CO: Westview Press, 1991, pp. 413-442. Rogers stresses that such codes often take the form of unarticulated assumptions and often operate on decisionmaking in a less conscious, more context-dependent, and often more insidious fashion than is commonly understood. Moreover, each decisionmaker within the same national leadership can prefer a different code.

12. *Ibid.*, pp. 416-417.

13. *Ibid.*, pp. 417-419.

14. *Ibid.*, pp. 419-421.

15. *Ibid.*, pp. 421.

16. Alexander George points out that the political and operational requirements of any crisis management approach or theory are not "requirements" in the strict sense of ". . . constituting necessary or sufficient conditions for successful management of war-threatening confrontations." However, whether these conditions are adequately met can affect the outcome of a crisis and a failure to adhere to them increases the likelihood of war. George, "Findings and Recommendations," pp. 560-561.

17. That is, the possibility of terminating a crisis short of war may depend on both sides having careful limits on both ends and means. The danger of escalation is obviously greater if both sides have ambitious objectives and means.

18. Scholars also point to other operational requirements that have little relationship to any particular crisis management strategy; i.e., they should be observed in any crisis. These requirements are largely concerned with the need to integrate military and diplomatic actions, and to maintain strong control over the deployment of military forces. For example, the movement of military forces and threats of force intended to signal resolve must be consistent with limited diplomatic objectives; political authorities must maintain informed control over military options and the selection and timing of military movements; and diplomatic-military options should be chosen that signal, or are consistent with, a desire to negotiate a way out of the crisis rather than to seek a military solution. Moreover, in order to avoid a shooting war, each participant must resist the temptation to seize opportunities during a conflict to inflict too severe a defeat on the other side; this might force the opponent into choosing between backing down or utilizing force in order to avoid a serious setback. See George, "A Provisional Theory of Crisis Management," pp. 23-25.

19. George, "Findings and Recommendations," pp. 553-554.

20. *Ibid.*

21. *Ibid.*, p. 554.

22. As Rogers states, "The mutual confidence of both sides that their basic security interests will be respected is the *sine qua non* of crisis management; without it, efforts at crisis management will, at best, be problematic." Rogers, pp. 420-421, 424.

23. George, "A Provisional Theory of Crisis Management," pp. 23-24.

24. When incentives to avoid war are weak, favorable opportunities for crisis management are quite limited, and relevant skills in short supply, then the likelihood of getting into a war is higher. Not only the ability but also the willingness of leaders to adhere to and satisfy the above political and operational requirements are significantly influenced by these variables. George, "Findings and Recommendations," p. 562.

25. Rogers, pp. 428-437.

26. As many scholars observe, multi-actor crises present special complexities and obstacles (see, for example, George, "Findings and Recommendations," p. 563). In particular, they pose difficulties in meeting the above political and operational requirements. A typical problem is the tension between the desire—at times, the necessity—for a major power to provide significant support to a regional ally and at the same time to not allow itself to be drawn into (or to inadvertently escalate) a dangerous, war-threatening confrontation. This is particularly relevant to the U.S.-China-Taiwan situation.

27. The following features are drawn from a wide range of studies of Chinese conflict behavior, interstate bargaining, crisis management, and strategic views

and beliefs (see Appendix II), as well as recent interviews and discussions with Chinese and American scholars conducted by the author in 2003 and 2004.

28. One should also point out that China has not anticipated always accurately the "worst case" outcomes in past crises. According to some Chinese observers, the Chinese leadership seriously underestimated the reaction of the Soviet Union following the Sino-Soviet border clashes of 1969. The Soviet response resulted in an unexpected escalation of the crisis, apparently involving the possibility of a major Soviet military strike designed to force China into negotiations, and perhaps ultimately to force Mao Zedong out of power. This sharp response eventually compelled Mao to deescalate the crisis and open talks with Moscow. And yet many Chinese analysts view Mao's retreat as a successful case of Chinese "escalation control!"

29. Despite an emphasis on clear signaling for deterrence purposes, China has shown some hesitation over the establishment of a direct line of communication with the United States through a hotline. This is apparently because: a) Chinese leaders apparently cannot respond to direct contact from the U.S. side until a consensus has been reached on how to proceed; and b) the Chinese are concerned that phone-based messages or signals might be misunderstood by either or both sides.

30. Interviews, Beijing and Washington, January-February 2004. See also Suisheng Zhao, eds., *Across the Taiwan Strait: Mainland China, Taiwan, and the 1995-1996 Crisis*, New York: Routledge, 1999; Andrew Scobell, "Show of Force: Chinese Soldiers, Statesmen, and the 1995-1996 Taiwan Strait Crisis," *Political Science Quarterly*, Vol. 115, No. 2, Summer 2000; Allen S. Whiting, "China's Use of Force, 1950-1996, and Taiwan," *International Security*, Vol. 26, No. 2, Fall 2001; Robert S. Ross, "The 1996 Taiwan Strait Crisis: Lessons for the United States, China, and Taiwan," *Security Dialogue*, Vol. 27, No. 4, December 1996; and Michael D. Swaine, "Chinese Decisionmaking Regarding Taiwan, 1979-2000," in David M. Lampton, ed., *The Making of Chinese Foreign and Security Policy in the Era of Reform, 1978-2000*, Stanford, CA: Stanford University Press, 2001, pp. 289-336.

31. Moreover, as indicated above and reiterated by Chinese observers, when facing a superior power, this willingness to use force derived, in turn, partly from a sense of vulnerability and weakness, and from the absence of other credible means to communicate resolve and to exert leverage in a crisis.

32. Perhaps one significant exception to this general pattern of behavior occurred during the Sino-Soviet military clash and subsequent crisis of 1969-70. At that time, China generally moved incrementally, in a seemingly tit-for-tat fashion.

33. That is, China's past tendency to regard crises in zero-sum terms, its sense of weakness compared with the United States (with its resulting resistance to appearing weak in a crisis), and its belief that many crises involve very high stakes could together reduce its concerns over falling into a commitment trap.

34. Some of the following 20 questions are posed by the author and some originate with other interested observers of China's crisis management behavior.

CHAPTER 3

THE TIANANMEN MASSACRE REAPPRAISED: PUBLIC PROTEST, URBAN WARFARE, AND THE PEOPLE'S LIBERATION ARMY

Larry M. Wortzel

The People's Liberation Army (PLA) of Communist China massed elements of more than a dozen infantry, armor, and airborne divisions around the city of Beijing in May and June 1989. Workers and students precipitated this massing of military force by demonstrating inside the city for more political expression and against nepotism and rampant corruption in the Chinese Communist Party (CCP). For weeks the CCP senior leadership was divided about how to handle the demonstrations. The political divisions in the highest ranks of the Party paralyzed decisionmaking and reflected ambivalence, even outright disagreement, in the PLA over the use of the military to suppress the unrest. A number of older, retired Communist Party members who fought to take control of China from the 1920s through 1949 emerged with comments and advice.[1] Many pushed for the suppression of the demonstrations by the PLA. Younger, reform-minded Party members coalesced around General Secretary Zhao Ziyang as he pushed for greater openness and economic liberalization.

This situation leading up to the Tiananmen Massacre is best characterized as a process of decisionmaking under stress. Initially the Communist Party Politburo Standing Committee (PBSC) was torn internally about how to respond, and the lack of response from student demonstrators put them under considerable stress. Later in the process, however, when it was clear that attempts to communicate with the students would not end the demonstrations, and the demonstrations spread throughout China on a larger scale, what had been a stressful situation became a genuine crisis.

In the end, the more orthodox Communists and other factions inside the Party who resisted any shift in the distribution of wealth and power won out. The students were suppressed with deadly force.

The disagreements over the pace of reform and the role of the Communist Party in the Chinese state that led to the slaughter of demonstrators and innocents alike in Beijing was really the extension of an earlier dispute about reform within the CCP that resulted in the dismissal of CCP General Secretary Hu Yaobang approximately 2 years earlier. It was Hu Yaobang's death on April 15, 1989, that threw the Communist Party leadership into turmoil and paralyzed inner-Party decisionmaking at the highest levels. The lesson from this chapter is that the consensus-related system of power among factions at the top of the Communist Party created, and will likely continue to create, paralysis in decisionmaking, thus delaying the ability of the CCP to react and potentially exacerbating any crisis.

A Party Divided.

Some basic framework to characterize relations among the senior leaders in China helps when analyzing political phenomena. As events played themselves out in China over the pace and depth of reform, the major political actors in the CCP formed into factions or "blocks" with different orientations on the issues. Some also would characterize the factional formation as a function of the perceived benefits to, or action on, certain "interest groups." For instance, the PLA was involved in the situation very early. As an institution, it had an interest in maintaining the power and authority of the CCP. It also had its own interests in seeing the demonstrations through to a peaceful conclusion if possible, because its leadership preferred not to be used against the Chinese populace. The public and state security offices functioned as another interest group in the process. What is clear, however, is that dividing the CCP leadership into merely "reformers" and "conservatives" was too simplistic an analytical perspective to capture the situation.

In his 1988 published analysis of the fall of Hu Yaobang, David Bachman argues that the Western observer must view the two ideological constructs central to the CCP dispute, "reformism" and "conservatism," as two poles of the Chinese political spectrum. But Bachman argues that analysts of Chinese politics should not conceive of all political actors as fully committed to groups or leaders at either pole. Thus factional formation and leadership is not merely a "two-line struggle"; coalition building is necessary.[2] It is the building of these coalitions, which for various reasons of personal and political interest span the poles of the ideological questions involved, that make up the active maneuvers in CCP political disputes. To deal in the Maoist concept of a two-line struggle of dialectical opposites around either pole of an issue simply is not how political life really works in the PRC, according to Bachman.

To illustrate his point, Bachman has suggested a useful characterization of the factions that coalesce around what are generally believed to be conservative political coalitions in China. According to Bachman, there are varying forms of conservatism in the CCP, of which he identifies six:[3]

1. *Financial Conservatives,* most of whom seek stability, predictability, and central control over fiscal and monetary policy. He sees the Ministries of Finance and the Central Banks as the major institutions that contribute Party members to this group, and identifies Chen Yun as the major figure in this faction.

2. *Planning Conservatives,* a group with a preference for the Stalinist command model of economic development with set economic targets, output quotas, and an emphasis on heavy industry.

3. *Moral Conservatives,* who focus for the most part on styles of behavior, a life of thrift and hard work that follows CCP mores. This group complains about lawlessness, decadence in the Party, overt sexuality, conspicuous consumption, and licentious dress. He identifies Deng Xiaoping, Zhou Enlai, and Liu Shaoqi as representatives of this group.

4. *Ideological Conservatives,* who see nationalism as important, dislike non-Party experts, fight against income differentials, and are generally opposed to reformers.

5. *Vested Interests,* who seek to retain the power of specific institutions such as the Public Security Bureau (PSB), the PLA, the ship-building industry, or the Ministry of State Security.

6. *Anti-Foreign Conservatives,* who support programs to keep foreigners separate from Chinese by providing certain

privileges for foreigners like special hotels, housing, and travel arrangements. These conservatives are nationalists with a "chip on their shoulder" who emphasize China's sense of being "wronged" by foreign powers in the past.

Thus, for varying reasons, different political actors in the Communist Party will coalesce around a conservative issue or cause. But they will not always agree on other issues. And some political actors, such as "moral conservatives," may be ready to throw out a centralized state planning system, threatening a key belief of "planning conservatives." The result of these differences creates a complicated and shifting set of political alliances. In the case of the Tiananmen Massacre, as this chapter will illustrate, the decision eventually came down to one set of choices—to use force to restore order or not—but the process of reaching that decision point was slow, and the internal arguments exacerbated the crisis. Moreover, the crisis over economic reform had brewed for some time, going back to the term of Communist Party General Secretary Hu Yaobang.

The Fall of Hu Yaobang and Inner-Party Struggle.

Failed political coalition management among reform-oriented senior Communist Party officials resulted in the forced resignation of Hu Yaobang as General Secretary of the CCP in January 1987.[4] This inner-Party struggle among coalitions and factions manifested itself in the dismissal of Hu Yaobang, but it played itself out over a longer period. The ideological differences over the speed of reform and arguments over the strength of the CCP grip over different aspects of the economy still were not resolved by 1989. They played themselves out during the period leading up to the Tiananmen Massacre, and the disruptions they caused in CCP political life directly contributed to the paralysis in decisionmaking at senior levels of the Party.

Complaints about Hu Yaobang among CCP elders and senior leaders were that he made serious mistakes in a number of areas: he resisted the campaign to oppose bourgeois liberalization; he opposed the Party's campaign against spiritual pollution; he shifted the focus of the campaign to eliminate opponents of reform by advocating the separation of party and state; his economic policies advocated high consumption; he supported a policy of rule by personality over rule by law; he was indiscreet about discussing inner-Party matters with foreign guests; and he did not properly consult other central leaders before making decisions.[5]

It is noteworthy that two ideological conservatives who opposed reform, Hu Qiaomu and Deng Liqun, were removed from power at the 13th Party Congress in October 1987. Even though removed from the center of power, these two actors were still part of a group of ideological conservatives who were influential in the inner-Party crisis of 1987, and in the disputes over reform in 1989. The group included Chen Yun, Li Xiannian, Peng Zhen, and Deng Yingchao (widow of Zhou Enlai). Both Hu Qiaomu and Deng Liqun emerged again in 1989 and were very critical of attempts to settle the demonstrations in any way that appeared to give in to the students.[6]

The Inner-Party Split.

What should have been apparent to those observing events in China at a much earlier point was the depth of the split within the Communist Party. One of the major points raised after the fact in the Party's own attempt to justify the Tiananmen Massacre was the April 16,1989, *dazibao* (large character poster) put up by students in the aftermath of Hu Yaobang's death. The text of this poster, "Commemorate Hu Yaobang, Protect Zhao Ziyang," is a clear indication that within the senior levels of the CCP, the split was deep and contentious.[7] More seriously, the general populace in Beijing was conscious of the split and, by putting up the poster, the more vocal clearly favored reform. Inside the Party, the tone of the poster only confirmed the relationship of the split between more orthodox elements of the CCP that objected to reforms and Zhao Ziyang's supporters who wanted to speed reform.

The death of Hu Yaobang, on April 15, 1989, catalyzed popular opinion in Beijing. Residents knew that the Party Center was debating reform and had been caught up in this debate since well before Hu's dismissal as Party General Secretary in 1987. Beijing's citizens used public expressions of grief over Hu's death to express their support for the reform process. This only created more turmoil in the CCP.

Zhao, with his policies already under attack by conservatives, used local Party organizations to mobilize mass support for his policies.[8] The initial campaign was subtle, focusing on universities and the cultural elite.

Indeed the discussion of how to respond to and influence the students was part of a continuous discussion inside the CCP. On April 24, at a PBSC Meeting, Li Ximing suggested that Party members and leaders at universities and from Municipal CCP Committees go out and "mingle with the students and do political work" on their thinking.[9]

This approach—trying to use the CCP members around Beijing both to influence the students and to get a sense of their thinking—continued throughout the confrontations on Tiananmen Square and the demonstrations. But by May 17, when it was clear that he might lose the day, Zhao went so far as to try to mobilize the military, workers in state-owned industry, and the government bureaucracy. More than a million persons marched in Beijing that day.[10]

If any single act confirmed the breadth and depth of the split in the CCP, it was Zhao Ziyang's statement to student leaders in Tiananmen Square on May 19, 2002, that that "we have come too late."[11] This tearful "confession," which should have been a signal to the occupiers of the Square that they were going to be dealt with in the harshest possible manner, came at 4 a.m. The decision to move troops into position around Beijing actually seems to have been made earlier, at a meeting on May 17.[12] Thus Zhao's statements on May 19 not only represented a certain breach of trust with his senior CCP colleagues, but also amounted to a real warning. The decision to move in the PLA and impose martial law followed a meeting of the elders of the CCP—Deng Xiaoping, Chen Yun, Li Xiannian, Peng Zhen, Deng Yingchao, Yang Zhangkun, Bo Yibo, and Wang Zhen. Even at this moment of drama, however, a few senior Party members seemed to vacillate between poles; or to use Bachman's formulation, to have different reasons to side with a "conservative" course of action.

For the military, Hong Xuezhi, Qin Jiwei, and Liu Huaqing agreed that using the PLA was the only viable alternative at this point. Yao Yilin, Hu Qili, and Li Peng supported the use of the military to impose martial law. Beijing mayor Chen Xitong, with no control over his own city and its Party apparatus impotent to stop or direct the protesting students and workers, also supported the decision.

Qiao Shi's role in the discussions was far more cautious, even ambivalent. As early as April 24, Qiao made it clear that maintaining "the growing democratic atmosphere in society" was an important goal in managing the demonstrations. However, in a PBSC meeting, Qiao also added that "unruly freedom or irresponsible freedom" could not be permitted.[13] Qiao noted that there had already been serious "beating, smashing, robbing, and burning" in Xi'an and Changsha, and wanted to prevent a repeat of that around China. As late as April 28, Qiao remained essentially tolerant of the students. He continued to advocate more dialogue with them while establishing clear political borderlines about behavior.[14] But by May 4, Qiao was very worried about how the students would act.

Before the imposition of martial law in Beijing on May 20, Zhao used the grass-roots party branches to support his own position. In every neighborhood, village, army organization, and work place in China, there is normally a blackboard or bulletin board run by the local Communist Party cell that relays the "Party line" for all to read and to follow. Whether at the PLA Air Force-controlled Nanyuan Missile Factory, at People's Armed Police (PAP) barracks, in the battalion cantonment areas of PLA units, or in urban neighborhoods in Beijing's suburbs, the message on these bulletin boards was the same. This author consistently observed communications from the CCP urging Party members and residents to support reform, support Zhao, support the students, and get out and demonstrate. This was really a last-ditch struggle for political survival by Zhao, who activated the grass-roots CCP infrastructure over which he had day-to-day control. But by using the Party infrastructure as he did, he sealed his own political fate of house arrest. He also confused the messages emanating from the Party center, contravened inner-Party discipline, and paralyzed the normal means for political dialogue.[15]

Looking back at April 17, at Beijing University, students in a number of dormitory buildings organized their peers and other students from Qinghua University to march to Tiananmen Square and lay a wreath for Hu Yaobang. Interestingly, at this point, the police and city authorities still were assisting the students and helping them bring off organized marches. Traffic was diverted to facilitate the student demonstrations, and they sang the "Internationale" in front of the Communist Party headquarters at *Zhongnanhai* after the wreath laying.[16] As noted earlier, CCP officials were encouraged to mingle with the students, both to sense their thinking and to do "ideological work." Their presence among the students may have confused the demonstrators about the nature of CCP support for their cause.

The active support of the PSB and the Party in making the marches flow smoothly only emboldened the students. Larger demonstrations took place on April 18, with some 10,000 students marching. They took over Tiananmen Square temporarily and again marched the one-half mile west to *Zhongnanhai*. This time, however, they did not disperse and stood their ground at the front gate, insisting on giving a letter to Party officials. These actions led to a confrontation between the students, the PSB, and the PAP in front of the main gate of the compound. To disperse another crowd on April 19, the police used truncheons, injuring a few of the students.

By April 21, the Beijing government and the CCP were already complaining that citizens seeking to hold the memorials for Hu Yaobang were disturbing the "social order" in the city.[17] Students had begun to camp on Tiananmen Square, and, while the government monitored them closely, neither the police nor the military took active action to remove them.

The CCP Signals Its Intent.

The April 26, 1989, issue of *People's Daily* (*Renmin Ribao*) reported a meeting in the Great Hall of the People attended by "cadre of the Party, Party members, workers, students, residents of areas around Beijing, and patriots." The student demonstrations were labeled a "turmoil" that the Party appealed for all to "resolutely and swiftly resolve."[18] The Politburo Central Committee, according to the article, accused a "small group of people" of using the event of Hu Yaobang's death and the students for the "reactionary call to overthrow the Communist Party." According to the article, 10,000 people attended this meeting. A similar meeting was held in Shanghai on the same day, the newspaper reported, this one attended by 40,000 Communist Party cadres from that city. In Shanghai the situation was called "unnatural" and was partially attributed to people coming into the city from other parts of the country (*wai di ren*). A parallel editorial was published the same day in *Renmin Ribao*, using the authoritative Communist Party bold, italicized type. This "social commentary" (*she lun*) once more accused a small group of reactionaries of continuing to use the students for their own reasons.[19] It also accused the group of organizing attacks on the Communist Party using the students, and of encouraging them to write "large character posters."

The day after the articles and the editorial, Xinhua Press ran another story that cited the incidents in front of *Zhongnanhai*, where students refused to disperse, as an "explosive situation" that was attributed to some students who intended all along to use the commemoration of Hu Yaobang's death as an excuse to attack the Communist Party.[20]

Indeed, while part of the Defense Attaché Office of the American Embassy at the time, this writer found notices on CCP bulletin boards inside military institutes urging Communist Party members to "get out in the street and demonstrate with the students in support of the students and workers, in support of Zhao Ziyang—protect reform and opening." This was quite a different message from that of April 17, which was for CCP members to circulate among the students and do ideological work. Clearly such efforts were part of the last-ditch effort by Zhao and his supporters to protect their positions. That Communist Party bulletin boards contained this line, and that it was repeated in a number of places—a military hospital, a logistics depot, and a military housing area—means that at some level the grassroots organizations of the Party were being used by Zhao as a tool to combat the more orthodox and older Party members.

Still, Zhao really had no senior party officials on his side. In a post-Tiananmen analysis by Beijing-based correspondent Nicholas Kristof, carried by *The New York Times Magazine* on November 12, 1989, those associated with Zhao Ziyang were Yan Mingfu, Bao Tong, Hu Qili, Rui Xingwen, and Yan Jiaqi. All of these men were Party officials who had risen to senior levels, but none were able to compete with the constellation of Communist elders, who had coalesced to remove Zhao and end the demonstrations. And the elders either were contemporaries to Deng Xiaoping or were politically allied to the more conservative wing of the Party. Kristof identified among them: Deng Xiaoping, Li Peng, Yang Shangkun, Chen Xitong, Li Ximing, Yao Yilin, Qiao Shi, Wan Li, Li Xiannian, and Chen Yun.

Using the typology suggested by David Bachman (pp. 57-58 of this chapter), it is possible to roughly categorize how those elders fit in to the conservative wing of the CCP from an ideological standpoint. Deng Xiaoping generally is thought of as one of China's moderates, pushing for economic reform, contact with the outside world, and modernization. Nevertheless, he was a moral conservative and a leader with strong ties to the vested interests of the Communist Party. Indeed, all of the elders identified in the Kristof article share this characteristic. Li Peng fits in with the group of "planning conservatives" and is also an ideological conservative. Yang Shangkun is another who is compatible with planning conservatives, is a moral conservative, and shares the vested interests of the Party as a major concern. Chen Xitong was an ideological conservative, a planning conservative, and is generally anti-foreign in orientation. Li Ximing and Yao Yilin are moral conservatives and planning conservatives. Qiao Shi is a moral conservative, as was Wan Li and Li Xiannian. However, when one looks at the videotape of Zhao Ziyang escorting Li Xiannian into meetings, one has to wonder whether Li was lucid at all. Assuming he was, the record in *The Tiananmen Papers* is that his major concern was the vested interest he had in the CCP. Yao Yilin, in addition to his emphasis on good central planning, also was very concerned in maintaining the vested interests of the CCP. Wan Li was almost completely out of the entire situation. He was summoned back from the United States, where he was on a visit, on short notice. Inside the U.S. Embassy, based on this writer's experience, some of us wondered whether he would defect rather than return to China. In the end he did return, and took very leglistic positions supporting the authority of the CCP to use force and enforce discipline. Still, many of us in the U.S. Embassy questioned whether his heart was in his statements. In the end, Wan was one of the most reform oriented elders in the Party. Chen Yun was probably the most senior and committed of the planning conservatives, was a moral conservative, and was generally anti-foreign in orientation.

By May 13, this split between the orthodox older CCP members and the Zhao reform oriented group became more apparent. *Tuanjie Bao*, a student publication in support of reform, noted that conservatives (the *baoshou pai*) were "struggling hard" against the moderate faction (the *wen he pai*).[21] The conservative charge was that Zhao and his allies were working with a "small group of conspirators" (*ji shao shu ren de yin mou*) to take over the Party and turn it into a bourgeois liberal group. Even as late as the day before the declaration of martial law, on May 19, 1989, major Communist Party-controlled organizations such as the State Economic Reform Institute, the State Council Research Center, and the Beijing Youth Research Center published letters urging the Party to "listen to the complaints of the students and citizens" about corruption and nepotism in China.[22]

As the likelihood of military involvement became more apparent, the self-understanding of the Chinese military affected PLA actions. The PLA bought into its own myth: that the PLA was a "people's army." At the start of the demonstrations the Army tried to stay out of the fray. The PLA leadership had sought hard to avoid being the primary instrument of force, hoping that the PSB was up to the task. In the end, the PLA was caught up in an inner-Party dispute for primacy in a power struggle over the speed and scope of reform in China.[23] Since 1978, when Deng Xiaoping institutionalized the reform of communism at the Third Plenum of the CCP 11th Central Committee, the nation was reforming, and private incentive systems increasingly were the norm. These reforms, which began as experiments in Deng's home province of Sichuan under then-governor Zhao Ziyang, stimulated initiative and economic growth. By 1989, with Zhao as CCP Chairman, the reforms also brought serious inflation. As a consequence, the conservative, orthodox Marxist elements in the Party wanted to slow, even reverse, the reforms.[24] This threatened Zhao's position, as well as that of other reform-minded central leaders.

Most studies on the Tiananmen Massacre start their narrative by examining the outpouring of grief and popular support for reform surrounding the death of Hu Yaobang.[25] The popular movement started earlier than that, however, as small private stalls and an "individual economy" began to flourish. By mid-1988, indigenous rock bands, theater groups, small private clubs, and restaurants had opened in most major Chinese cities, more or less institutionalizing cultural and economic reform. Centralized, orthodox Stalinist planning was being changed, and the change undermined the sinecure of CCP bureaucrats, some who had occupied their positions since 1949, and others who had risen during the upheavals of the Cultural Revolution. Thus, from the national level to the local level, whether in the cities or the countryside, there was an entrenched cadre of CCP operatives opposed to reform.

It was this inner-Party fight that paralyzed the political processes of China. Under normal circumstances, the "democratic centralism" of the Communist Party provided a means by which disagreements could be surfaced, conveyed to the Party center, and dealt with.[26] The CCP might modify a policy based on this feedback or remain firm in its approach, but there was a mechanism that at least permitted the CCP to sample inner-Party opinion, if not public attitudes, make a collective decision at the top reacting to opinion, and communicate that decision to cadre and citizens. However, it was not a quick process. As the system functioned, once a decision had been made, local Party committees conveyed the new line in meetings. On more important decisions, the *People's Daily* and local, but CCP-controlled newspapers, might publish an editorial conveying or reinforcing the line. The general population shut up and followed the new line on pain of severe criticism or banishment to some labor farm. But the reforms were so popular and the impact on entrenched cadre so strong that the normal "feedback loops" were paralyzed.[27] Moreover, once Party members began to get mixed messages from the Party center, the situation became more confused. June Teufel Dreyer, in her 1989 assessment of the inner-Party arguments over reform, put it this way:

> The Army's sluggish response and the apparent disunity in its ranks can be explained, at least in part, by the fact that its condition largely mirrors that of the Chinese polity as a whole. In the political sphere, factional disputes among party leaders have traditionally been reflected in similar cleavages within the military. The events of the spring thus involved not only a power struggle between different leadership factions, but also

a parallel struggle between conservative and reformist factions in the military . . . the army suffers from the same woes that afflict civilians.[28]

With the mounting pressure to resign now coming from his colleagues on the CPP Central Committee, the Politburo, and the military, Zhao saw the handwriting on the wall—his resignation had to come. Still, he attempted to marshal last-minute support from those in the populace that favored reform.[29]

The Hong Kong and Western press picked up the factional splits in the CCP fairly quickly. But the disagreement in the Party over how to react to, and end, the student demonstrations was the focus of attention. The *South China Morning Post* set out the factional alignments pushing Deng Xiaoping toward a hard-line crackdown on the students as including Chen Yun, Bo Yibo, Song Renqiong, Hu Qiaomu, Deng Liqun, Li Peng, Yao Yilin, and Wang Renzhi.[30]

South China Morning Post correspondent Willy Wo-Lop Lam best summarized the situation in what he reported as a conversation between Deng Xiaoping and Zhao Ziyang. According to Lam, in a heated discussion, Zhao said to Deng Xiaoping that the Party must support reform because "I [Zhao] have the people on my side."[31] In Lam's version of this conversation, Deng supposedly responded, "You have nothing! I have the Army." Willy Lam probably was not in a position to know what actually may have been said in a Zhao-Deng dispute; those exact words may not have been used. Lam lacked the access and placement to obtain such direct quotes from any conversation that may have taken place at such an elevated level within the CCP. Nevertheless, Lam's column captured the essence of the dispute within the Party. It was this dispute that was a major factor in the eventual reaction by the PLA to the demonstrations.[32]

Gathering Discontent.

The initial public gatherings in Beijing began on April 16, 1989, with students gathering on the campus of Beijing University.[33] Hu Yaobang was a popular figure as the Party general secretary, but was removed from CCP positions in 1987 in the aftermath of student demonstrations supporting further liberalization in China. Hu sped reform, but his haste broke a tacit, inner-Party understanding that the older generation of revolutionary leaders, who at Deng Xiaoping's urging moved to the "second line" in 1986, would continue to guide the country from "behind the curtain." Although Deng had anointed Hu as his successor, Hu's association with speedier reform challenged the elders. For this reason, Hu was removed from office and replaced by Zhao Ziyang.

Even without this challenge to inner-Party discipline, other problems plagued the country. Inflation was high, and corruption and the use of personal influence and "connections" (*guanxi*) circumvented the political, legal, and social system, effectively paralyzing society. Official profiteering from the mixed government quota-market system (*guan-dao*) corrupted commerce and made it impossible to work within the established system.[34] There were no fixed rules or laws; only access and influence produced legal or business decisions. Most seriously, however, the people who were profiting were senior CCP and military leaders. The children of Deng, Chen Yun, and the revolutionary veterans were placed in critical positions in military-run companies. Sons, daughters, nieces, and nephews of active duty, retired, and even deceased generals and marshals (the "party princelings" or *tai zi dang*) corrupted the system for their own benefit, and everyone knew it—workers, students, lower level government functionaries, and intellectuals.[35]

Even the forces of order (or repression) in the cities, like the PLA General Staff Department-controlled PAP, were part of the problem. The PAP ran networks of underground bars and nightclubs in Beijing and other major cities. The PAP (along with PSB colleagues) controlled prostitution. The PAP (and units of the PLA) owned the trucks used to smuggle illegal goods.[36] Military unit commanders knew full well that a retrenchment by the Party would kill the golden goose of market reform whose eggs were used to line their pockets. It is not surprising, therefore, that when the government began to move military units to suppress student demonstrations as early as April 15 (to coincide with Hu Yaobang's death), the PLA and PAP leaders were torn between personal self-interest and Party loyalty.

The networks of corruption complicated the reaction. Units of the PLA were involved deeply in hotel management in Beijing. Thus, although the Group Armies of the PLA were far from the capital, the children of the army, division, and regimental commanders were embedded in business in the city. The PAP, which should have been the first line of defense in restoring order, lived in the city among the populace. PAP junior officers and senior cadre married girls from the neighborhoods of Beijing. PAP soldiers, when they stood duty, flirted with the girls and played with the children in the neighborhood. When finally called on to help restore order, many PAP and PLA organizations instead facilitated the marches. These were the last people who would "break heads" to end a crisis.

On April 17, when several thousand university students marched toward Tiananmen Square seeking a reassessment of Hu Yaobang by the Party, the PAP (and PSB) facilitated their progress through the traffic of Beijing. Of course, seeing that it might strengthen his own weak position, Zhao Ziyang encouraged the students and the PAP through Party grass-roots organizations. By April 22, 200,000 students demonstrated, again quietly assisted along by the PAP and PSB. This happened not only in Beijing, but in other cities as well, including Xi'an, Changsha, Chengde, and even the hometowns of the 27th and 38th Group Armies, Shijiazhuang and Baoding.[37]

Normal political dialogue and feedback in the Communist Party depended on a "reading" of signals relaying the final Party position under "democratic centralism."[38] The April 26 edition of *People's Daily* contained an editorial that labeled the student movement as an "anti-party" action that was "anti-socialist" and was a form of "turmoil." This editorial used a series of "code words" that relayed the concerns of senior leaders, who had been severely repressed during the Cultural Revolution, that the limits of forbearance had been reached. Under normal circumstances, the average student or citizen would have "read the tea leaves" and toed the Party line. But these were not normal times, and the political signals were confused by the inner-Party battle.

It was at this time that Zhao Ziyang and his pro-liberalization Party colleagues realized they were in serious trouble. Their reaction was to encourage workers, government bureaucrats, and industrial employees to demonstrate using the grass-roots Party organs and bulletin boards. If one entered a PLA or PAP compound, the Party bulletin boards told the Army to support Zhao, support the students, and encourage reform. Army commanders heard one thing from their children, another from their orthodox political commissars, and still another message from those in the Party struggling for liberalization and survival. The result was chaos.

When ordered to move toward Beijing, some military commanders and soldiers reportedly feigned illness. The most notorious case of this involved the commander of the 38th Group Army, Xu Qinxian.[39] Xu's case came to the direct attention of the members of the PBSC because his father was General Xu Haidong. Other military commanders, in explaining the deployments, emphasized to their soldiers that, as a "people's army" the PLA should obey the Party but take no action to hurt the populace. The PLA's involvement, according to this line of reasoning was designed to stabilize the situation. Meanwhile, the demonstrations continued. Almost a million people marched in Beijing on April 27; 100,000 people in Shanghai on May 2; 300,000 people in Beijing on May 4; 10,000 students and writers in Beijing on May 10.[40]

On May 13, with PLA units surrounding the city but staying on its outskirts, the students declared a hunger strike. For the next 3 days 100,000 people a day demonstrated all around Beijing, some in front of PLA units. The army units were torn in their loyalties. Students and city residents bombarded them with good will, providing water and food at times. The Zhao Ziyang wing of the Party, by activating grass-roots party branches, encouraged them to support reform, and the orthodox political commissars in the units told them the demonstrations were an anti-socialist mass movement. When Soviet President Mikhail Gorbachev visited Beijing on May 15, ceremonies had to be shifted, and the CCP clearly had no control over the city, causing a major loss of face for China's leaders. The climax came on May 17, when a million people, including military organizations, marched in Beijing.

This writer, serving as a military attaché and moving around the city, felt the probability of the use of deadly force by the Army to settle the demonstrations became clear as early as May 6, when Zhao called for more openness in an address to political workers. Zhao later disclosed that it was Deng, not himself, that was making all the important decisions in the Party.[41] This judgment was reinforced on May 18, when Li Peng met with student leaders in the Great Hall of the People and was insulted on national television by student leader Wuer Kaixi. So senior a leader as Li Peng, representing the most conservative elements of the CCP, could not have this challenge go unpunished.

Friction, Weather, and a Critical Missed Opportunity.

Clausewitz uses the term "friction" in war to describe unplanned events or circumstances that complicate operations or plans. Had Wuer Kaixi been a good "Confucian" (or Communist) and listened to his elder and not insulted Li Peng, violence might have been avoided. But he did not act according to script, introducing "friction" into the calculations of the central leadership of the CCP. Weather and judgment played a factor, too. On the night of May 17, the weather was unseasonably terrible for Beijing—chilly and rainy. The students were wet, cold, and hungry. Those on Tiananmen Square were becoming hypothermic, and probably could have been peacefully herded off the square on the morning of the 18th if the Army had acted at dawn. Instead, hundreds of buses from the city transportation bureau miraculously showed up on Tiananmen Square, giving the protestors relief from the weather and subtle reassurance that someone in the CCP supported their cause, thus creating more friction. Interestingly, it was the night before that, according to accounts in *The Tiananmen Papers*, when the decision was made to impose martial law. In a bureaucracy like China's, this sort of thing does not happen unless some official of the Communist Party makes it happen. It took the actions of a number of different organizations, all controlled in one way or the other by the CCP, to make those buses appear. The Beijing City government owned and controlled the buses, and Communist Party leaders controlled both the government and the bus bureau. The PSB had to clear routes for the buses, and the PSB, of course, is controlled by a CCP Committee. The most likely explanation is that the appearance of the buses, which took so much coordination, was the hand of moderates and reformers supporting the Zhao wing of the CCP. The fate of the demonstrators and Zhao's supporters was sealed. Moreover, on May 18, the 5th day of the hunger strike by the students on Tiananmen Square, after the buses showed up to provide shelter on the square, military people, some with their units identified on banners, joined demonstrators on the streets of Beijing.[42]

Turning again to the PSC meeting of May 17, the collective decision there was to try to "expose the tiny minority (of students) who are agitating and creating turmoil," while at the same time to "soften confrontations." This was to be accomplished somehow at the same time that the CCP managed to "protect the patriotism and enthusiasm of the students and the broad masses."[43] In the debate that took place at the PBSC meeting, Qiao Shi agreed to the deployment of the PLA, but expressed his hope that force would not have to be used. He advised taking advantage of the way that the declaration of martial law might stabilize parts of Beijing, but also taking advantage of any pause it might bring about by having CCP branches encourage parents to get their children out of the square. Qiao's admonition to the PBSC was "we do not want to shed blood."[44]

The Army Mobilizes.

Premier Li Peng and Beijing Mayor Chen Xitong declared martial law on May 19, to take effect on May 20. Once more, Qiao Shi agreed to the declaration of martial law, but expressed hope that no force would have to be used. On the evening of May 19, elements of the PLA attempted to enter the city from all four cardinal directions. The units identified as involved in the initial deployment were elements of the following PLA Group Armies: 24, 27, 28, 38, 63, 65, the Beijing Garrison Command, 39, 40, 54, and 67 (the latter four from outside Beijing Military Region).[45]

Workers and citizens rallied to the student cause. The mere appearance of troops and the declaration of martial law did not have the dramatic effect of cooling the demonstrations that the CCP elders and PBSC thought it would. Ordinary residents who realized that continued demonstrations could mean a chance to end corruption joined together with the student movement to block the PLA and come to the defense of the city. One very poignant moment on May 20 demonstrates how the people of Beijing reacted to martial law. Having been moving around the city all day taking stock of events, this attaché was dirty, hot, thirsty, and tired when approaching a roadblock designed to stop the PLA manned by workers on the second ring road in Beijing, near the Bell Tower. Two young men wore PLA ammunition bandoleers over their shoulders filled with one-liter Beijing beer bottles. When offered money for a bottle of beer, a young man replied, "These are filled with gasoline; when the PLA comes after the students, they'll see what 'People's War' really means. We'll give it to them." Two weeks later that is exactly what happened.

On the eve of Children's Day, May 31 (Children's Day was June 1), an italicized letter appeared in the main position above the fold of *Peoples' Daily* from Deng Yingchao. This was the prototypical grandmother of the Communist Party, the widow of Zhou Enlai. She was probably the most senior of the CCP officials who were also revolutionary veterans and was, indeed, not only a symbolic grandmother to the young students in the square, but really a grandmother. In the article she made it clear that June 1 was children's day, and a grandmother should be able to take her grandchildren to Tiananmen Square without the interference of demonstrations—the students should clear the Square. Despite the agreement with the "Party elders" on the use of the PLA to stabilize the situation, this was a final public warning to the students and other demonstrators. Her appeals were ignored.

The PLA Prepares: A Classic Military Operation.

It was clear to China's leaders that the police and PAP were ineffective in controlling the crowds, if not disloyal, and that large portions of the PLA leadership sympathized with the demonstrators. Moreover, additional forces were pouring into marshaling areas outside Beijing. Nanyaun Airfield south of Beijing, for example, got troops in from the 15th Airborne Army in Kaifeng to reinforce units already there from the 54th Group Army. The 12-14 divisions of different PLA group armies surrounding Beijing were pulled back, away from the city, into military assembly areas. Sequestered and isolated from the populace in those encampments, the uneducated, rural infantrymen and tankers of the Army were told that the city was full of a combination of good, but confused citizens and "counter-revolutionary enemies of the people." This use of language was a reversion to the most virulent Maoist terminology associated with class warfare. One could hear it on loudspeakers at night when close to the troop assembly areas.

At dawn on June 3, a regiment of the 196th Infantry Division from Tianjin attempted to run into Tiananmen Square unarmed. They were turned back by citizens and students. This seems to have been the final catalyst for the issuance of a final order to the PLA. The "Basic Order" issued to the troops was that they were to cross the line of departure outside their assembly areas at 9 on the evening of June 3. The initial units were ordered to arrive at Tiananmen Square by 1 a.m. on June 4, and the square was to be cleared no later than 6 a.m.. The PLA was to permit no delays to interfere with the accomplishment of this mission, and was ordered to clear blockages "using any and all self-defensive means required."[46] Thus the order to clear the Square authorized the use of deadly force.

At about 4:00 on the afternoon of June 3, a full armor division of the PLA left its assembly area near the town of Tongxian, some 40 kilometers east of Beijing, and formed up along the Beijing-Tianjin Highway in assault positions at what must have been their designated line of departure. The division, like other PLA units around the city, had undergone almost 2 weeks of "reeducation" designed to convince the soldiers that the demonstrators were counter-revolutionary criminals. The troops were armed. Behind the tanks were armored personnel carriers, followed by trucks full of infantrymen. To the south, in the vicinity of Nanyuan Airfield, troops of the 54th Group Army and 15th Airborne Army began to form for the assault. Farmers, sensing that the attack on the "counter-revolutionaries" was coming, lined up tractors and backhoes along the roads between Nanyuan and the Temple of Heaven to block the PLA. In the western and northern parts of the city, the same things happened. The stage was set for a phased military operation to gain control of Beijing.

People's War Against the People's Army.

From the southern approach outside Nanyuan Airfield, assembled PLA units began their approach march to the city at about 2 a.m. The lead elements were helmeted PAP, not soldiers, who moved in front of the troops, trying to disperse the urban defenders with nightsticks. They were hopelessly ineffective. Behind them, the PLA tanks cut through roadblocks of city buses, trucks, and tractors like butter. As the PLA got closer to the center of the city, resistance increased. Young urban workers, most of whom either had army or militia training, really did conduct a "People's War." After tanks passed through a road block, the urban fighters often used the steel bars from road dividers to disable them, as well as the armored personnel carriers, by breaking their tracks. Once a combat vehicle was disabled in this manner, street fighters swarmed over the vehicle, covering the engine intake with blankets on which they had poured gasoline or diesel fuel. When the blankets were ignited, the PLA crews were forced to exit the vehicles, after which they were beaten by the people at the roadblocks and pelted with rocks or Molotov cocktails (the Beijing beer bottles full of gas). Terribly burned soldiers ran among crowds, their clothing in flames. This was clearly a tactic rehearsed and even practiced among the demonstrators, since it was used in the same way in separate places around the city. By this time, having seen their fellow soldiers killed, the troops were scared. Political commissars had told them for 2 weeks that there were "counter-revolutionary criminals" in the city. Predictably, the troops reacted by opening fire. Meanwhile, the disabled armored vehicles blocked the progress of the trucks and reinforcing troops. A similar scene happened west of Tiananmen Square, along Changan Boulevard, in the "Battle of Muxidi."[47]

By the time troops neared Tiananmen Square, they were scared and angry. The political reeducation at the hands of the CCP commissars had come true. There really were "bad elements" inside the city. Literally any resident of Beijing foolish enough to be on the street was a potential "criminal," and the troops shot at them indiscriminately. Still, this was a military operation, and some discipline was obvious; not a single foreign reporter or diplomat was shot.

June 4.

As dawn broke on June 4, 1989, the streets of Beijing were littered with brass shell casings from the PLA AK-47s and from the machine guns on PLA tanks and armored personnel carriers. The bodies of dead students and workers filled hospitals.[48] Burned-out buses, trucks, and armored personnel carriers lined the streets.[49] Blood, bandages, and the hard-tack biscuits carried by soldiers were ground into the concrete.[50] A column of smoke rose from the center of the city around Tiananmen Square, and the air was filled with the smells of burning flesh mixed with the odor of petroleum. PLA helicopters ferried back and forth between Tiananmen and airfields in Tongxian, 40 kilometers east of Beijing, Shahe to the north of the city, Nanyuan to the south, and Xiyuan (nestled at the base of the Western Hills near the PLA General Staff Department (GSD) underground command complex). Hanging from a pedestrian cross-walk on Chongwenmen Street, south of Tiananmen, was the disemboweled body of a PLA soldier burned so badly that his skin was browned and his limbs drawn up like an overcooked chicken.[51] Half-a-mile away near Qianmen, the corpse of another disemboweled soldier was hanging from a burned out bus.[52] A sign placed around his neck by local residents said, "This man murdered an old woman and her granddaughter." A city that was vibrant and pulsating with new ideas, rock music, private businesses, restaurants, and bars was transformed overnight into a scene from hell.

The PLA fractured its own self-created myth of being a "people's army" when it turned on the residents of Beijing.[53] But this was not the first time it had done so. From the time of the land-reform movement of the late 1940s and early 1950s forward, the PLA had shown its willingness to carry out the orders of the CCP central leadership, inflicting pain and death on those who were identified as "enemies of the Party."[54] Since the height of the Cultural Revolution (1966-69), however, when the PLA was called on to restore civil order, it really had not been used enmasse against the Chinese people. Of course, throughout history, the Chinese military has always been used to keep the people in line, but Tiananmen hit a generation of Chinese who had bought into the myth that it was more a "people's army" than a "party army."[55]

The Death Toll.

A *New York Times* reporter, Nicholas Kristof, after visiting a few hospitals, arbitrarily set the death toll in Beijing at 900. Truth be told,

Kristof had no idea of the actual toll. The PLA and the CCP set the death toll at 200, of which they claimed that only 36 were students. Twenty-three members of the PLA and PAP were killed (10 PLA and 13 PAP).[56] Early in the morning of June 4, a discreet source of information to the Defense Attaché Office in the Chinese Red Cross set the toll at 2,600, but for 3 more days Beijing was full of gunfire. A PLA defector in Hong Kong in 1996 set the toll at 3,700.[57] This may be nearer the truth, but no one knows. One thing is certain, however, despite the claims of Communist Party spokesman Yuan Mu in 1989 and Minister of Defense Chi Haotian at the U.S. National Defense University in 1996, people did die on Tiananmen Square. How many is unclear. Officers of the U.S. Embassy watched people have their heads blown apart by PLA bullets on the Square. Journalists watched people crushed by vehicles. A week later, the monuments on the Square were still stained with blood and chipped by bullets. The monuments were also marred from armored vehicles driving on them. One must ask the question, if no students were left on the Square, why was there blood, why were there bullet holes, and why did armored vehicles drive onto the monuments?[58]

Scaring Away Foreigners.

Immediately after gaining control of Tiananmen, PLA units began a methodical campaign to capture demonstrators and eliminate resistance around the city. In some parts of Beijing on July 4 and 5, shooting continued sporadically, even after dark. Rumors began to spread, repeated to the press by Western diplomats, that units of the PLA were going to fight each other. In actuality, the PLA was acting in a coordinated manner. Supply vehicles from the units rumored to be fighting each other could be seen refueling and drawing food and ammunition together at the same PLA supply points.

Both the Army and the CCP were tired of the constant, watchful eyes of the foreign press and diplomats. On June 7, the PLA opened fire on hotels and diplomatic housing in the embassy district. The military claimed that a sniper had fired on a PLA unit from within the diplomatic housing compound at Jianguomenwai. This was not true. This writer was warned in advance and given the exact hours of the shooting incident in enough detail that he was told what floors of what buildings in the diplomatic compound to avoid. The PLA action was part of a carefully planned event to scare the foreigners out of the city, and, for the most part, it worked. Embassies and businesses withdrew most foreign staffs and their families.

Civil-Military Relations Repaired.

The Tiananmen Massacre was a tragedy of monumental proportions. The fact that a massacre took place, and so many lives were lost, can be attributed only to the vacillation inside the CCP and ultimately the Communist Party's bungling paralysis. The demonstrations and unrest were resolved reluctantly only by brutal urban combat. The leaders of the PLA worked hard to avoid the Army's involvement. And clearly, at least from the objections raised initially by PSC members like Qiao Shi, senior CCP officials would have preferred to handle the demonstrations peacefully. But when ordered to do so, the Party-controlled PLA used deadly force against the Chinese people. Chinese armies always have done this to preserve a dynasty, perhaps they always will.

In the aftermath of the Tiananmen Massacre, entire infantry divisions of PLA converted to PAP, simply changing uniforms to provide a strong, reliable force in the city. High school and college students were taken to the countryside for military training in the summer. This familiarized them with the Army and humanized the soldiers in the eyes of young men and women. In the intervening years, when China reduced the size of the PLA, the PAP grew proportionately. PLA junior officers and squad leaders were sent into intermediate schools and high schools in Chinese cities to assist in physical education classes to teach rudimentary hand-to-hand combat and close-order drill. The students liked the change in school routine, and parents, products of the Cultural Revolution, felt that their sons and daughters were more disciplined at home and in school. Civil-military relations improved. Perhaps the most significant event that made the PLA more of a "People's Army" once more was the 1998 flooding in China. Here the Army reacted swiftly to assist the populace, regaining respect.

Conclusion.

David Bachman's characterization of coalition formation inside the CCP holds up well with time. For different reasons, senior and "elder" members of the Party coalesced around conservative positions over the time of the crisis. Their primary goal was the maintenance of power and authority for the Party when it was faced with real calls for reform. Some joined against the moves for reform to preserve their own power, some to maintain their networks of influence, nepotism, and corruption, and some perhaps out of dedication to a cause. But until they coalesced into a cohesive body and while there was serious disagreement within the CCP, there was also paralysis. Ultimately, even those who counseled against force, like Qiao Shi, eventually agreed to the "Basic Order" and the use of deadly force. One faction manipulated the populace of the city, if not the country, in one direction, while another faction pulled in another. The normal signals that sent people to their deaths in the Cultural Revolution, like editorials in Party newspapers, had little effect. And the result was paralysis.

The lessons from the Tiananmen Massacre are that, in the final analysis, the CCP will work for its own survival as an institution. And the Communist Party can probably count on the PLA to do its bidding. Another important lesson of the vacillation in advance of the Tiananmen Massacre is that serious inner party struggle will result in paralysis every time that the nominal leadership of the Party, to include its General Secretary and the head of the Central Military Commission, have to go back to a "board of trustees" of older, more senior CCP members for a decision. This sort of consensus leadership by committee does not lend itself to decisionmaking under stress, let alone crisis management and resolution. Without some legitimate, institutionalized constitutional form of authority, this leadership format likely will exacerbate future crises rather than help to resolve them. Whether responding to domestic crisis or international crisis, the CCP and the PLA's inability to follow an established crisis management procedure will exacerbate the situation.

ENDNOTES - CHAPTER 3

1. In its September 1989 analysis of events leading up to Tiananmen Massacre, the Directorate of Intelligence of the Central Intelligence Agency CIA) described the "old guard" of Communist Party elders as "a small group of party elders—most in their 80s and 90s—who have criticized Deng's [Deng Xiaoping's] economic reform program as straying too far from orthodox Marxism-Leninism." Included in this group by the CIA are Chen Yun, Peng Zhen, Bo Yibo, Wang Zhen, and Li Xiannian. See Directorate of Intelligence, Central Intelligence Research Paper, *The Road to Tiananmen Crackdown: An Analytic Chronology of Chinese Leadership Decisionmaking*, EA-89-10030, Washington, DC: Central Intelligence Agency, September 1989, p. vi. Authorized for public release in a redacted version in March 2000. This author thinks the CIA analysis is too simplistic and finds the work by David Bachman a more nuanced understanding of how some of these Party elders supported some elements of modernization, but differed with Deng elsewhere (see footnote 2).

2. David Bachman, "Varieties of Chinese Conservatism and the Fall of Hu Yaobang," *Journal of Northeast Asian Studies*, Spring 1988, pp. 24, 44.

3. *Ibid.*, pp. 26-35.

4. *Ibid.*, pp. 22-46.

5. *Ibid.*, p. 40.

6. Zhang Liang, Compiler, and Andrew J. Nathan and Perry Link, eds., *Tiananmen Papers: The Chinese Leadership's Decision to Use Force Against Their own People-In Their Own Words*, New York: Public Affairs Press, 2001, p. 4. Hereafter cited as *Tiananmen Papers*.

7. Editorial Board, Beijing Publishing House, *The Truth About the Beijing Turmoil*, Beijing: Beijing Publishing House, 2nd Printing, August 1990, pp. 7-8.

8. "Leaders Still Locked in Power Conflict," *South China Morning Post*, June 4, 1989, p. 3.

9. *Tiananmen Papers*, p. 58.

10. *Ibid.*, pp. 193-196.

11. Michael Oksenberg, Lawrence R. Sullivan, and Marc Lambert, eds., *Beijing Spring 1989, Confrontation and Conflict: The Basic Documents*, Armonk, NY: M. E Sharpe, 1990, pp. 288-291.

12. *Tiananmen Papers*, p. 189.

13. *Ibid.*, p. 60.

14. *Ibid.*, pp. 87, 104.

15. "Yang Speech Details Moves Against Zhao," *South China Morning Post*, May 30, 1989, p. 11.

16. *New York Times*, April, 18, 1989.

17. "Women zenyang daonian Hu Yaobang Tongzhi?" ("How should we hold a memorial service for Comrade Hu Yaobang?") *Renmin Ribao*, April 21, 1989. This article is part of the collection catalogued in *The China Democracy Movement and Tiananmen Incident: Annotated Catalog of the UCLA Archives, 1989-1993*, UCLA Asian Pacific Monograph Series, compiled by Jian Ding and Elaine Yee-Man Cha, Los Angeles: University of California at Los Angeles UCLA), 1999, catalogued as CC1012. The author thanks the library staff at the UCLA East Asian Library for assistance and access to the collection.

18. *Renmin Ribao*, April 26, 1989.

19. *Ibid.*

20. "Beijing jin ri chuxian yixie sishi erfei chuanwen yongyuan fangmian zhiqing renshi pilu shishi zhenxiang" ("Rumors are Spreading in Beijing: Knowledgeable Officials are Revealing the Truth"), *Renmin Ribao*, April 28, 1989, catalogued at UCLA as CC1043.

21. See the 39 pages of student flyers in the UCLA Archives catalogued as AC1006, dated between May 13, 1989, and June 4, 1989.

22. *Guanyu shiju de liu dian shengming* (Six Point Declaration on the Current Situation), May 19, 1989, Catalogued in the UCLA Archives as AC1008.

23. Ben Tierney, "Old Long March Guards Under Pressure of Change," *The Hong Kong Standard*, June 3, 1989, p.8.

24. Bachman, p. 22-46; "Attacks on Zhao Renewed," *Foreign Broadcast Information Service-China* (FBIS-CHI)-89-250, January 2, 1990; "Article Cites Discord Among CPC Leaders," FBIS-CHI-89-250, January 2, 1990.

25. Michael Fathers and Andrew Higgins, *Tiananmen: The Rape of Peking*, London: *The Independent*, 1989; *Massacre in Beijing: China's Struggle for* Democracy, New York: Time Books, Inc., 1989; David and Peter Turley, *Beijing Spring*, New York: Stewart, Tabori and Chang, 1989; *China in Crisis: The Role of the Military*, Surrey, UK: Jane's Information Group, 1989; Theodore Hua and John Li, *Tiananmen Square Spring 1989: A Chronology of the Chinese Democracy Movement*, Berkeley: University of California Press, 1992; *June Four: A Chronicle of the Chinese Democratic Uprising*, Hong Kong: Ming Pao Publishing, 1989; Timothy Brook, *Quelling the People: The Military Suppression of the Beijing Democracy Movement*, New York: Oxford University Press, 1992. Andrew Scobell, *China's Use of Military Force: Beyond the Great Wall and the Long March*, Cambridge: Cambridge University Press, 2003, chapter 7.

26. Stanley Karnow, *Mao and China: From Revolution to Revolution*, New York: The Viking Press, 1972, pp. 162, 230.

27. Maria Hsia Chang, "The Meaning of Tianamen Incident," *Global Affairs*, Fall 1989, pp. 12-35.

28. June Teufel Dreyer, "China After Tiananmen: The Role of the Military," *World Policy Journal*, Fall 1989, p. 648.

29. *Tiananmen Papers*, pp. 179-181, 184-190, 257-273. A few words about the authenticity and accuracy of *Tiananmen Papers*: Former National Security Council director for China and CIA officer Robert Suettinger says, "the authenticity of the document, in my opinion, is questionable. It is just a bit too neat, too convenient, and too sympathetic to Zhao (Ziyang)." See Suettinger's *Beyond Tiananmen*, footnote 44. This author agrees with Suettinger's assessment. Much of book seems to be a review of events reported in the press and most resembles a reconstruction of events from memory after the fact, perhaps assisted by a review of news accounts.

30. *South China Morning Post*, May 28, 1989, catalogued in UCLA Archives as CE1002.

31. Willy Wo-Lap Lam, "Deng May Agree on Compromise," *South China Morning Post*, May 29, 1989, p. 4.

32. Willy Wo-Lap Lam, "Soldiers Prepare to Move Into Beijing," *South China Morning Post*, May 31, 1989, p. 3.

33. Fathers and Higgins, *Tiananmen: The Rape of Beijing*, pp. 4-11.

34. Gene D. Tracey, "The People's Liberation Army As 'Bad Iron,'" *Asian Defence Journal*, November 1989, p. 31-42.

35. Guocang Huan, "The Roots of the Political Crisis," *World Policy Journal*, Fall 1989, pp. 609-620.

36. James Mulvenon, *Soldiers of Fortune: The Rise and Fall of the Chinese Military-Business Complex, 1978-1998*, Armonk, NY: M. E. Sharpe, 2001, pp. 57-89.

37. *Tiananmen Papers*.

38. Mao Tse-Tung, "On the People's Democratic Dictatorship," in *Selected Works of Mao Tse-Tung*, Vol. IV, Peking: Foreign Languages Press, 1975, pp. 411-424.

39. *Tiananmen Papers*, pp. 213, 219, 239-240.

40. UCLA Tiananmen Archives AC1003, AC1006, AC1008, and CC1030.

41. Directorate for Intelligence, CIA; *The Road to Tiananmen Crackdown*, pp. 5-6.

42. This author observed a number of groups marching with banners identifying the marchers as from military logistics organizations. Also see *Tiananmen Papers*, p. 217.

43. *Ibid.*, p. 188.

44. *Ibid.*, quote from p. 175; 210.

45. *Ibid.*, p. 212. It is important to remember that PLA units do not make deployments in such numbers in a mere 2- or 3-day period. Despite the record in *Tiananmen Papers* that a deployment decision was made on May 17, it is likely that the General Staff Department issued a mobilization order far earlier, perhaps as early as the time of the April 26 *People's Daily* editorial.

46. *Tiananmen Papers,* p. 370.

47. *Ibid.,* pp. 372-382.

48. Up to 2,500 Killed in Carnage at Tiananmen," *The Hong Kong Standard,* June 5, 1989, p. 1.

49. "Protestors Set Tanks on Fire," *International Herald Tribune,* June 5, 1989, p. 1.
50. "Tanks Ignored Screams," *The Hong-Kong Standard,* June 5, 1989, p. 2.

51. "Corpses, Tanks Paraded in Capital," *The Hong-Kong Standard,* June 5, 1989, p. 5.

52. The author of this chapter personally observed these events and was able to interview the local residents about the circumstances under which the citizens of these districts attacked the soldiers.

53. Bernard E. Trainor, "An Army's Agony," *The New York Times,* June 19, 1989, p. A10; June Teufel Dreyer, "The People's Liberation Army and the Power Struggle of 1989," *Problems of Communism,* September-October 1989, pp. 41-48; Kerry Dumbaugh, Shirley Kan, and Robert Sutter, "China's Prospects After Tiananmen Square: Current Conditions, Future Scenarios, and A Survey of Expert Opinion," *CRS Report for Congress,* January 15, 1991.

54. Andrew G. Walder, "The Political Sociology of the Beijing Upheaval of 1989," *Problems of Communism,* September-October 1989, p. 30-40.

55. Andrew J. Nathan, "Chinese Democracy in 1989: Continuity and Change," *Problems of Communism,* September-October 1989, pp. 16-29. Scobell, *China's Use of Military Force,* p. 165.

56. Oksenberg, Sullivan, and Lambert, pp. 363-376.

57. Information provided to author by Chinese defector in Hong Kong.

58. For further reading on the subject, see also Michael Fathers and Andrew Higgins, *Tiananmen: The Rape of Peking,* London: *The Independent,* 1989; *Massacre in Beijing: China's Struggle for Democracy,* New York: Time Books, Inc., 1989; David and Peter Turley, *Beijing Spring,* New York: Stewart, Tabori, and Chang, 1989; Paul Beaver, ed., *China in Crisis: The Role of the Military,* Surrey, UK: *Jane's Information Group,* 1989; Theodore Hua and John Li, *Tiananmen Square Spring 1989: A Chronology of the Chinese Democracy Movement,* Berkeley: University of California Press, 1992; *June Four: A Chronicle of the Chinese Democratic Uprising,* Hong Kong: Ming Pao Publishing, 1989; Timothy Brook, *Quelling the People: The Military Suppression of the Beijing Democracy Movement,* New York: Oxford University Press, 1992; Robert L. Suettinger, *Beyond Tiananmen: The Politics of U.S.-China Relations, 1989-2000;* Washington, DC: Brookings Institution Press, 2003; James A. Lilley with Jeff Lilley, *China Hands: Nine Decades of Adventure, Espionage, and Diplomacy in Asia,* New York: Public Affairs Press, 2004.

CHAPTER 4

SARS 2002-2003:
A CASE STUDY IN CRISIS MANAGEMENT[1]

Susan M. Puska

Introduction,

> There have been as many plagues as wars in history; yet always plagues and wars take people equally by surprise.

Albert Camus[2]

Seldom does a domestic health emergency spin out of control the way the Severe Acute Respiratory Syndrome (SARS) crisis did in China during the early months of 2003, threatening global health and economic stability. After over 5 months of denial, as information of the spread of the disease to Beijing was exposed, growing external pressure forced Chinese leaders to shift into action. The Chinese Communist Party (CCP), headed by General Secretary Hu Jintao since the November 2002 16th Party Congress, when the epidemic emerged, was forced to dramatically shift its SARS response strategy between late March and early April 2003, as foreign confidence that the leadership had the situation under control evaporated. Fearing economic and international implications, the CCP leadership initiated aggressive and highly visible actions in the fight against SARS by mid-April 2003.

The CCP's strategy of denial and deception had served it well from November 2002 to February 2003, as SARS spread from Guangdong Province into Hong Kong and radiated out internationally. The uncertainties of the origins, nature, and vector of the disease aided the Beijing leadership to delay its release of information, while obscuring facts that were known. As SARS spread, official dissemination of inaccurate and incomplete information to an increasingly skeptical international media and officials became less effective.

If the Chinese Communist Party and State leaders hoped the SARS problem would go away with minimal consequences if they simply ignored it, by late March their strategy unraveled as scrutiny became more intense and critical and the effects of SARS in Beijing became publicly known. On March 31, the *Wall Street Journal* published a commentary entitled "Quarantine China," that bluntly criticized Chinese officials for withholding information about the spread of SARS. It called for a temporary ban on travel to infected areas, including China, and the screening and quarantine of persons who had been exposed to SARS. WHO issued a travel advisory the following day.

Between late March and April 1, international events scheduled for Beijing, such as the International Ice Hockey Federation 2003 Women's World Championship, were cancelled or postponed. By April 1, word that the China Economic 2003 Summit, scheduled to start on April 14 in Beijing, was postponed, raised concerns about the economic repercussions of SARS and heralded another flood of postponements and cancellations.

On April 6, Finnish national Pekka Aro became the first foreigner to die in China of SARS. Although Beijing health officials tried to obscure his case, blaming his infection on foreign exposure, the efficacy of continued deception weakened further. It collapsed by April 9, the same day that *Time Magazine* published leaked information of falsification of SARS statistics that Jiang Yanyong, a military doctor at the 301 military hospital, had provided the China Central Television and Hong Kong-based Phoenix television on April 4.

The Party and government were joined in their belated fight by the mobilization of the People's Liberation Army (PLA), particularly health and anti-chemical assets, under the direction of Central Military Commission (CMC) Chairman Jiang Zemin. Working together, the Party, government, and military curbed further spread of the epidemic by June 24, 2003, when the World Health Organization (WHO) removed its travel advisory for Beijing. As a result, China's leadership largely regained international confidence and enjoyed praise for its "decisive" action against SARS. Even though China's eventual response to the crisis showed how national resources could be concentrated for a short period of time, what is more telling is the protracted failure to respond effectively to the epidemic during the 5 months between mid-November 2002 and mid-April 2003.

This chapter will focus on how the CCP's crisis response methodology allowed SARS to spread within China and internationally. It raises questions about the CCP's ability to handle future crises, especially public welfare problems.[3] As the Party's key guarantor of stability and power, the PLA's mixed record and short-comings in civil-military cooperation will also be discussed.

The SARS crisis illustrates how the CCP's priorities have become so intertwined with the Party's own survival and maintaining a monopoly on power that the Party leadership from the bottom to the top often cannot balance public interest against their own self-interests. Further, the SARS case illustrates that the CCP has also become dependent on foreign investment and trade to underwrite its legitimacy that it will delay decisions and conceal information in order to protect foreign economic interests, rather than promote the public welfare. This latter point was driven home when the CCP only decided to take action after SARS had radiated out internationally from China, and information about the rate of infection in Beijing had been leaked to the international community. But, in the end, foreign pressure and scrutiny can still encourage Beijing to take positive action.

The SARS Crisis and China's Response—A View from Beijing.

Rumors of a previously unknown and deadly fever first surfaced in November 2002. When reports of incidents of an "atypical pneumonia" occurring in Southern China (Guangdong Province, later identified as Fushan City)[4] reached Beijing. Southern China is not only known as an economic power-house for China's modernization, it is also one of the world's disease hot spots. Scientists have determined that most new flu strains originate in Southern China, where humans and animals live in close proximity.[5] After an initial flurry of rumors, the mysterious disease seemed to disappear until early 2003, when it resurfaced in Vietnam and Hong Kong. On March 12, 2003, the WHO issued a global alert for "atypical pneumonia" cases in Guangdong Province, Hong Kong, China, and Vietnam.[6]

Appendix I presents a timeline of events related to SARS from November 2002, when the first known cases occurred in Fushan City, Guangdong Province, until late 2003/early 2004, when a second outbreak of SARS occured in China. Information on specific actions taken or not taken is spotty for the period between mid-November and the end of the year 2002. By early January 2003, however, there is evidence that at least some in the PLA in Beijing already were aware of the seriousness of the disease.[7] Guangdong Provincial authorities officially knew of the deadly disease at least by the beginning of January. They issued guidance in January that was ambiguous enough to avoid disrupting the New Year holiday. By late January, Guangdong leaders officially reported the situation to Beijing, but underreported the rate of infection and recommended Beijing impose a media blackout. In early February, the CCP propaganda organization issued guidelines for reporting SARS that directed all should stress that the situation is under control.

News of the SARS epidemic leaked out via the internet during early February after a SARS patient was treated at the Guangdong No. 2 Hospital, where he infected hospital staff. The Guangdong Party Secretary, Zhang Dejiang, continued to enforce a complete media blackout until February 11, when Guangdong health officials convened a press conference. They reported that only 305 people had been infected, five of whom had died, but they insisted the disease was under control. (This number was subsequently adjusted to 792 cases, and 31 deaths at this time.)[8] At the press conference, the head of the Guangdong Health Department, Huang Qingdao, further obscured the situation when he implied that the disease could be prevented, even cured, and that Guangdong had taken the right steps to control it. The reality, however, was that critical information about the rate of infection and the effectiveness of basic hygiene measures was not collected and shared even within Guangdong Province. The Guangzhou Air Force Hospital, for example, did not have any spread of infection, mainly by employing basic infection procedures. But the hospital did not share what it had experienced, which may have allowed the disease to spread unchecked and into the capital, Guangzhou, and beyond.[9]

On February 12, the Nanfang Military Hospital in Guangdong Province was the first to perform an autopsy on a SARS patient. The hospital concluded that the patient's death was caused by a virus-caused pneumonia. It distributed tissue samples from the SARS patient to the Guangdong CDC, Guangzhou CDC, and the No. 8 People's Hospital (which provided the corpse).

The following day, the Academy of Military Medical Sciences (AMMS), which has oversight of possible biological and chemical attacks, sent two researchers, with the approval of the General Logistics Department (GLD) and the Ministry of Health (MOH), to collect a specimen. Even though the Guangzhou Military Region and Nanfang Military Hospital provided assistance to the researchers, they were only permitted to collect a thumb-sized lung tissue sample, some serum samples, and a few drops of saliva.

By mid-February, two lung tissues from the Nanfang autopsy were provided to the Beijing Center for Disease Control (CDC). These samples were divided between Hong Tao, the CDC's chief virologist, and two other researchers. All three conducted separate studies. Hong's conclusion that chlamydia, a common sexually transmitted bacterium that is not particularly deadly, was the pathogen for the atypical pneumonia became the officially accepted theory.

Zhu Qingyu, one of the AMMS scientists who obtained a small sample from the Nanfang Hospital, detected a distinct halo of spikes, which indicated coronavirus, on February 20. On February 26, Zhu concluded that a new coronavirus was linked to the atypical pneumonia, but his findings, which challenged Hong's chlamydia theory, were rejected. Between March 21 and 22, Zhu conducted more tests on SARS samples in Beijing which provided even more evidence that a new coronavirus was linked to atypical pneumonia. He reported his findings to the GLD and MOH for approval. Chinese research of atypical pneumonia was inhibited by the officially accepted, but erroneous, chlamydia theory from February until early April, and the lack of cooperation within and between military and civilian researchers.[10] The overarching policy issued by the CCP propaganda department, however, ensured that information was concealed or underreported, while the threat of the disease was downplayed. This state of affairs lasted through March, when the 10th National People's Party Congress was held in Beijing.

By mid-April, however, confidence in the Central Government's ability to handle the crisis dissipated when the Minister of Health was exposed for deliberately concealing information on the rate of infection in Beijing. Health concerns within the foreign community in Beijing fueled panic, as news of official deception leaked to the Western Press during April 4-9, 2003. Text messages had been circulating within Chinese circles since early February, but this was the first official acknowledgement of the seriousness of the situation and the presence of the disease in the national capital. The key source of the leak was Jiang Yanyong, a 72-year old military surgeon, who worked at the 301 PLA Hospital in Beijing. Reacting to Minister of Health Zhang Wenkang's April 2, 2003, press conference claim that there were only 12 SARS cases in Beijing, Jiang passed information directly to China Central Television and Hong Kong-based Phoenix television on April 4. Neither released the information. The information then passed to the Western press. Jiang revealed that Beijing 301 military hospital alone had more than 100 SARS cases and six deaths.[11] Once the story broke and the government acknowledged a coverup by mid-April, it raised even more questions about how much more the government might be concealing about the extent of the epidemic in Beijing and elsewhere.

In Beijing, which would eventually experience the largest known[12] outbreak of the disease during 2003,[13] reports of atypical pneumonia were largely overshadowed in March by the 10th National People's Congress (NPC) that began on March 5, and the Iraq War, which began soon after the conclusion of the NPC. Ironically, on the same day the NPC started, a jewelry saleswoman traveled from Guangzhou to Beijing with atypical pneumonia symptoms and introduced SARS into the city.[14]

Although WHO issued a global alert in Beijing on March 12, 2003,[15] it was not until after word spread of the death of a 52-year-old Finnish national, Pekka Aro, from SARS on April 6, that foreigners began a hasty evacuation from Beijing and other cities in China.[16] By late April, Chinese citizens also fled from Beijing without any checks or restraints until roadblocks (official and local) were set up to restrict movement into and out of the city.

Official, commercial, and tourist visits to China were delayed or cancelled, although almost empty flights continued for several weeks. Travelers who did venture out of Beijing from April to June were subject to frequent temperature checks (high fever being a possible, but not foolproof indicator of SARS infection, but one that was embraced with draconian gusto by May). The PLA Armament

Engineering Academy, and perhaps other units, adapted military technology to take the temperature of moving crowds at 30 meters. These walking detectors were placed at airports and train stations throughout China. As far as Qinghai Province, which had no reported cases of SARS, body temperatures were scanned at airport, railroad stations, and road side check points using infrared scanning devices. Small infrared devices or conventional thermometers were also used at hotels and other locations to record body temperatures.

Some foreign companies temporarily suspended business or relied on local staffs to maintain limited operations. Western Embassies and businesses authorized nonessential personnel and families to leave the country for months until the WHO travel advisory was lifted in late June. During April and May, the crowds of Beijing dramatically thinned out as foreigners departed and locals fled, and those that remained stayed inside their homes or, in the case of an estimated 30,000 persons, were under quarantine.

During the May holiday, normally one of the busiest times of the year, Beijing streets were empty, while people stayed home watching an endless stream of special television programming. For those caught on the inside during a quarantine of a work compound or college campus, life was greatly circumscribed, although some continued to move freely in and out of compounds. Children continued their studies at home on the computer and television. Military compounds restricted personnel, although necessities continued to enter compounds through side doors. With the exception of taxi drivers, who lamented the poor business, drivers in Beijing enjoyed the emptied streets, rather than the regular choke of cars and trucks on Beijing's ring roads and streets. Those who ventured out not only could move freely, unusual bargains were to be had from shopkeepers eager to sell.

Ubiquitous white cotton masks hid the faces of many Chinese and a few foreigners. The cotton was a useless defense against a microscopic virus, but it provided a sense of reassurance, just as vinegar, incense, and other old folk remedies provided the illusion of safety against SARS. For a brief period, even the sound of hacking and spitting stopped, as people became more attentive to public hygiene. Outdoor cafes and restaurants ballooned around the city of Beijing as people concluded they were safer outside in the air.

Although some in Beijing had feared the SARS epidemic would spread further with the exodus of residents and floating population during late April, the worst fears never materialized. Even though SARS was contagious and sometimes deadly, it proved far less contagious than flu. The nightmare scenario of a pandemic flu, which could surpass the 1918 global flu, when 50 million died (or as high as 100 million, by some estimates) out of a population of 1.8 billion, did not materialize.[17]

In retrospect, although significant, the number of deaths and infections from SARS proved to be relatively modest. Worldwide the total number infected was 8,098. Of these, 774 people died. Although about 30 countries were infected with SARS, the hardest areas hit were China (5,327 infected, 349 dead), Hong Kong Special Administrative Region (1,755 infected, 299 dead), Canada (251 infected, 43 dead), and Singapore (238 infected, 33 dead).[18] Even allowing for possible underreporting in Beijing and elsewhere in China, the health damage from SARS was relatively minor. Economic damage resulting from the crisis proved recoverable for China within a matter of months.[19]

As a health crisis management case, even though SARS did not result in a pandemic, it should be viewed as a warning to China and the world of the necessity of early detection and response to new diseases. In China, SARS was able to develop unchecked from November 2002 to April 2003, largely due to China's inability to effectively respond to the disease. The crisis shows how the politics of deceit in the midst of a major health crisis with international implications can squander precious time, which can permit a new virus to spread and evolve with deadly consequences. Although the leadership under Hu and Wen eventually did a better job managing the disease and cooperating internationally, once official underreporting was exposed by Doctor Jiang in mid-April, there is also mixed evidence that China cannot be expected to respond proactively in the future in terms of surveillance, detection, and control of newly developing diseases, without international pressure and monitoring.

Publicly, WHO was extremely patient with China during the 2002-2003 SARS crisis, praising China for its cooperation even before mid-April[20] when Chinese officials admitted to underreporting SARS cases. Once China shifted to a political style mobilization campaign against SARS during mid-April, the implications of how the crisis had been mishandled between November 2002 and April 2003 became clearer. In early April 2003, David L. Heymann, a WHO official testifying before the U.S. Congress said, "We feel that China is taking the measures now [that] they can . . . If these measures had been taken in November, perhaps the disease would not have spread."[21]

While the world health community concluded that greater openness and action at the inception of a new disease is essential, China's leadership from Beijing down to the local (provincial and below) governments appears to have reached a different conclusion. As the point of origin for SARS, Guangdong Province did not suffer any serious sanctions for failing to adequately report the disease to the Beijing beginning in November 2002. Neither Guangdong nor Beijing expressed any concern or contrition that SARS had become a global event because of a failure in China's own crisis management.

The CCP's approach to problem solving, which demands secretiveness and deception, will likely even continue, as China's belated response to the reoccurrence of SARS during late 2003-early 2004 demonstrated. Without strong incentives to change, and lacking checks and balances on a one-party system, the world can expect a slow response from China during the next new disease. If that disease proves to be a more highly contagious new influenza, for example, China could inadvertently play the key role in spreading a pandemic.

SECTION 1. CRISIS MANAGEMENT—CCP STYLE

'[I]t is fine not to tell the public' because [I] am not legally required to do so.

Guangdong Health Chief, February 11, 2003[22]

To analyze the CCP's handling of the 2002-03 SARS epidemic, it is necessary to review the context of the period. Throughout this time, the overriding backdrop for the Chinese leadership was preparation for the leadership transition that would be announced during the 16th Party Congress in November 2002 and the 10th National People's Congress in March 2003. The Party leadership, as well as China watchers, was obsessed with the wrangling and horse trading that takes place prior to these major meetings from the local level up to the Central government. These meetings were significant, especially since they seemed to indicate a major generational shift between the leadership of Jiang Zemin to Hu Jintao. All were preoccupied with who would move up, who would move out, and how Jiang's legacy (encapsulated in the Three Represents) would be portrayed.

Major changes during an election year in a Western system can also be disruptive, but in China, the transition, which does not involve any change in the Party in power, is more protracted and secretive. Decisions, compromises, wins, and defeats are fought out largely behind the scenes among Party members, who are directed not to discuss these matters outside of party channels[23] until decisions have been reached and the unified line is ready for public viewing.

Given that this major leadership transition was the backdrop for SARS, it is little wonder that the Chinese government was slow to react to the SARS crisis. But this should not be exaggerated, and, in fact, it only reinforces how brittle China's one-party system can be. The CCP's routine behavioral, organizational, and informational crisis management characterized China's response to SARS. Lacking institutionalized checks and balances and a watchdog free press, the CCP is largely unable to police itself to eliminate even the endemic corruption within the Party, which has been an ongoing struggle for much of China's recent history. A crisis, such as SARS, threatens the delicate balance the Party maintains to perpetuate its own legitimacy, while also balancing broad reaching challenges that are posed by a large, diverse country undergoing one of the most extensive modernization and economic development transformations in modern history.

In addition to the pending leadership transition, the CCP faced a number of crises with international implications during 2001-03. Among these, the April 2001 EP-3 crisis, the September 11, 2001 (9/11), attack on America and subsequent invasion of Afghanistan, and the SARS crisis during late 2002–mid 2003 are particularly noteworthy. All three of these crises provided China's leadership with opportunities to promote or degrade China's regional and international national interests. They also provided situations where individual leaders, mainly Jiang Zemin and Hu Jintao, could demonstrate their abilities during a crisis.

During each of these crises before SARS, CCP leaders relied on routine and predictable characteristics for crisis management. The overriding priority in any crisis for the CCP is to preserve its power and reputation as the essential vanguard political party, regardless of the costs. An extreme example of this was during the June 4, 1989 incident, when the Chinese leadership used the PLA to suppress peaceful protest. But the CCP has been equally obsessed with self-preservation even when the threat is relatively minor, as they have demonstrated with the relentless pursuit of Falugong, even if its heavy-handed action tarnishes its international reputation.

Fearful of any potential loss of power, Party leadership is less motivated to analyze even a health crisis, such as SARS, based on empirical information, and they are less motivated to collect objective data that would assist decisionmaking, as well as ensuring synergy between government, Party, and military national assets.

Consequently, protection of the Party leads to a defense mentality when dealing with a crisis. The leadership will "circle the wagons" by delaying acknowledgement of an event and concealing and/or distorting relevant information, while the collective Party leadership negotiates its position. Even a leader of Jiang Zemin's stature is not usually free to make a unilateral decision based on his position and authority within the Party leadership. His self-proclaimed initiative to extend his condolences to the United States without Party approval after watching the attacks on New York and Washington, DC, on September 11, 2001, evidences how unusual it is for noncollective decisionmaking in the post-Deng Xiaoping political environment.

Internal negotiations to reach a Party decision can be so protracted; they inevitably result in delays even in openly acknowledging a crisis. In the case of a medical emergency such as SARS presented, such paralysis through negotiation can prove disastrous to domestic and international health interests.

The collective leadership may even be unable to take any action at all during an initial phase of assessment and negotiation to reach the Party line. Consequently, those who are authorized to speak for the Party during this phase will often deny and distort facts to stall for time. If information about the incident has been made public, the Party will generally blame others—foreigners and/or domestic troublemakers are most usually to blame, regardless of the situation.

In the late 1980s, for example, when cases of acquired immune deficiency syndrome (AIDS) were exposed in China, it was an article of faith that only Westerners could have AIDS. When it became known that Chinese were infected, the focus shifted to blaming Westerners for infecting Chinese nationals, rather than in determining the extent of the infection and how to minimize the spread of the disease within China. In another more recent case, the EP-3 incident in April 2001, there was never any question that the Chinese pilot's error could have contributed to the accident.

Once the Party line is reached, the Party and government propaganda departments work collectively to disseminate the official story. Information that contradicts the Party line is ignored, denied, or ruthlessly suppressed.

In the case of the SARS crisis in April 2003, PLA doctor Jiang Yanyong directly challenged the official line on SARS when he questioned the Minister of Health's deliberate deception and underreporting. Once he went public to the Western media, the Party was presented with a difficulty, but at the time they had little choice but to temporarily ignore the characterization of Jiang as a national hero of conscience.

As happens frequently during a crisis, the CCP and/or individual leaders will take the opportunity of a crisis to promote an advantage. For example, the Chinese negotiators sought to use the EP-3 incident as a means to promote recognition of China's broad sovereignty claims, which extend well beyond international limits, while it negotiated the release of the crew and later the aircraft with the United States. During April 2003, SARS presented Hu and Wen with an opportunity to consolidate their power,[24] and may have ultimately helped encourage Jiang Zemin to step down from his position as CMC Chairman in September 2003. But the SARS crisis was also an international public relations opportunity for them, for by doing anything, they were largely perceived as new and open leaders that the West could work with.

In general, the CCP's crisis management style is time-consuming and may be disconnected from the "facts," which could be counterproductive to handling a health emergency. In the case of SARS, China's delay in handling the epidemic and cooperating with the international community created the conditions for the disease to spread within China, including to the capital, as well as to some 30 countries.

The lack of oversight and transparency in the CCP's crisis management style encourages a high degree of groupthink and inflexibility when confronting a crisis. Once a decision has been made, it is very difficult to acknowledge an error by the Party. Scapegoats, such as the minister of health and the mayor of Beijing, who were both sacked in April, must be found, even if they are largely symbolically punished.

Hu's and Wen's consultation with non-Party public relations and medical experts in mid-April to assess the extent of the health crisis and determine what actions were required at that point to regain international confidence and contain the disease may have demonstrated the shortcomings of a politically-motivated crisis management system. But it is not likely to result in dramatic political change in China, since CCP cohesion and discipline are essential to its continued rule, and even Hu and Wen seek to improve the Party rather than replace it or add political competition.

Since alternative views are not encouraged and may be harshly sanctioned, CCP problem solving is highly limited. Adjustments after the party line has been reached are extremely difficult without a triggering event, such as Dr. Jiang's leaking of information and international pressure, which helped prompt Hu and Wen to take seemingly bold and open initiatives to confront the crisis in mid-April.

Role of the Military—An Assessment.[25]

As a sub-element of the Party,[26] the PLA's contribution to the crisis and its resolution is an interesting case within a case. Information about SARS infection at the PLA 301 Hospital were certainly concealed until Doctor Jiang leaked the information, but it is not entirely clear whether or not the PLA independently concealed information from the MOH. It seems more likely that the MOH, which had coordinated with the GLD on SARS since at least February, was aware that the SARS infection had reached the PLA 301 Hospital in March, particularly since the patient was a civilian. In fact, the opening of military hospitals to civilians with money to pay for care complicates the PLA's ability to control information about military health care. More likely, the PLA was enforcing Party propaganda guidelines to conceal information prior to mid-April. Nonetheless, the rate of infection among military personnel was withheld from public reporting in Beijing until mid-May, when WHO released statistics that about 150-160 military personnel were infected with SARS. Guangdong military hospitals, however, had agreed to report SARS infection rates to WHO during mid-April.[27] Although the PLA can be criticized for withholding information on the number of military personnel that were infected, it is understandable that they would feel the need to conceal this information since it directly reflected on personnel readiness at the time. What is less understandable, however, is why Guangdong military units, such as the Guangzhou Air Force Hospital, did not disseminate critical health information to other military units. In particular, although the hospital suffered no cases of staff infection, the unit apparently did not share this information through the chain of command. If it had, SARS infections of staff in the PLA 301 and 302 hospitals could have been avoided.[28]

The AMMS' difficulty in getting samples[29] from the first autopsy from the Nanfang Military Hospital in early February showed how problematic civil-military cooperation was prior to the Party order in April. Even another military unit with top level support from the GLD and MOH, as well as the Guangzhou Military Region, could only obtain a limited amount for its research. Once the academy's own researcher, Zhu Qingyu, made an initial discovery of coronavirus that contradicted the Chinese CDC theory, the academy was prevented from putting its theory forward. If they had, China would likely have been first to identify the pathogen for SARS and this information might have encouraged China's leadership to react more quickly to the threat.

The most important role the military played in the crisis came with full mobilization when military technology, primarily from biochemical capabilities, was adapted to civilian use. The PLA produced numerous protective devices, which were often developed in collaboration with civilian companies. Leading the PLA in these efforts were units and research activities of the GLD, which has purview over military medical assets, and the General Armament Department, which included anti-chemical capability. Several military regions also contributed to the effort.

Lessons for the Future.

Being proactive is crucial. A reactive approach costs lives.

Barbara Wahl, President
Ontario Nurses Association[30]

The relative transparency that Hu Jintao and Wen Jiabao displayed during the SARS crisis mobilization was largely reactive and reflected political pragmatism in the face of increasing international pressure. Domestically, it provided an opportunity to positively promote the new leadership team.[31] Although China's efforts to contain SARS within Beijing and the rest of the country after mid-April can be praised for intense action in a short period of time, the fact remains that inaction and deception between November 2002 and March 2003 resulted in the spread of the disease.

China's inaction and concealment of the next SARS outbreak during late 2003 and 2004[32] showed that the CCP's natural tendency to conceal information and delay crisis response remains largely intact. While China's response to the spring 2004 outbreak was better and indicated that international pressure can encourage a faster response to SARS, questions still remain about China's ability to handle a serious health crisis. Even with the best of intentions and consolidated power, Hu Jintao would face an uphill battle to try to change the crisis management character of the CCP from the top down, assuming he even wants to do this.

Transparency, openness, and cooperation with the international community will likely continue to be carefully balanced against the Party's own imperatives of survival and dominance. In the case of SARS, Party interests ultimately converged with international demands for greater accountability and action. In the future, however, the Party leadership cannot be expected to risk CCP dominance, regardless of the cost to public health or other issues. The tendency to conceal and deceive in order to maintain stability while preserving foreign interests will likely remain the central guidepost for CCP decisionmaking during a crisis. Hence, like Party corruption, the pattern of crisis management displayed during the 2002-03 SARS crisis will likely continue to characterize how China responds in the future. These limitations argue for intensified international cooperation and monitoring of China's health crisis management to encourage China to modernize its health care system to curb the spread of pandemic disease.

APPENDIX I

SARS 2002-03 CRISIS TIMELINE[1] WITH 2004 ADDENDUM

Phase One – "Atypical Pneumonia" begins in Guangdong Province

2002

November 2002

November 8-14 - 16th Party Congress convenes in Beijing.

November 16 - First case occurs in Fushan (Foshan) City, Guangdong Province. At least two patients become infected with an atypical pneumonia of unknown origin. Similar cases are soon reported in five Guangdong cities. A 35-year-old chef working in Shenzhen is transferred to the Heyuan People's Hospital, Heyuan City, where he infects at least 11 people.

November 2003-January 2004 - An unusual pneumonia spreads through the Pearl River Delta. State-run newspapers strongly deny any outbreak. WHO asks Chinese Health Ministry to comment on reports that health workers are becoming infected. The Health Ministry claims it is a minor outbreak of influenza B.

December 2002

Mid-December – Two SARS patients seek treatment in Heyuan City. They infect eight medical workers.

Late December – Following the small outbreak in Heyuan City, Guangdong Province imposes a local news blackout.

2003

January 2003

Early January – Exaggerated rumors spread about the death of three medical workers in Heyuan. Many people line up to buy antibiotics. Local officials try to calm the public by denying the existence of the disease in local newspapers and meetings.

A retired Chinese military logistics officer privately warns an American diplomat to avoid hospitals in Beijing because of a "deadly disease" that is spreading.

January 1 – On or about January 1, the Guangdong authorities learn of the deadly disease. Guangdong Provincial health team goes to Heyuan City to investigate cases at the People's Hospital.

January 2 – A second breakout occurs in Zhongshan, infecting over 12 patients and hospital workers.

January 21 – On or about the 21st, Guangdong Province issues a vague warning to provincial hospitals and health officials regarding the seriousness of atypical pneumonia, but fails to emphasize how infectious the disease is and identify what steps should be taken. Many officials did not begin to act on the warning until about February 7, after the Chinese New Year holiday.

Late January – Leading Guangdong epidemiologists survey the outbreak and conclude that the disease is an unusual form of pneumonia. Although the provincial public health bureau leaders have been aware of the outbreak, they do not report it to Beijing until late January as the disease continues to spread rapidly. They report only 600 cases, while 600 more cases remain hidden as "suspected" cases. They recommend Beijing initiate a media blackout on any news of the epidemic to preempt international criticism and maintain domestic stability.

Phase Two - Infection Spreads through Guangdong and Beyond (Hong Kong, Vietnam, Canada, Singapore)

February 2003

Early February – Based on the recommendation of Guangdong authorities, the Propaganda Department of the Chinese Communist Party directs that reports of SARS should follow specified guidelines and should emphasize that the situation is under control. Guangdong authorities use the directive to tighten control over media that openly discuss the disease. They issue up to three prohibition statements per day and crack down on the more outspoken media.

February 3 – A 40-year old man checked into the No. 2 Hospital in Guangzhou with symptoms of atypical pneumonia. He infects members of the hospital staff.

February 7 – *Southern Daily*, the official CCP paper in Guangdong Province, reports that the province has notified Beijing of the outbreak. *Washington Post* reports that the *Southern Daily* circumvented a news ban by obtaining the permission of the Provincial Governor, Huang Huahua, reportedly allied to Hu Jintao.

February 8 – Text message sent to mobile phones in Guangzhou: "There is a fatal flu in Guangzhou." Message will be resent 40 million times that day, 41 million times the next day, and 45 million times on the third day after the original message, according to *Southern Weekend* paper, published in Guangzhou. During February 8-10, as rumors spread, people in Guangdong rush to buy *banlangeng* (a Chinese medicine to treat colds), vinegar (believed to kill germs), antibiotics, masks, and salt. Prices of these products soar.

February 9 – Beijing reported to have sent an investigation team headed by Deputy Minister Ma Xiaowei to Guangdong Province.

February 10 – Rumors of outbreak appear on *Teachers.net.com* website when an American fourth-grade teacher receives an e-mail from a Guangzhou colleague, asking: "Have you heard of the terrible sickness in my city?" She passes the e-mail to a former Navy physician, who is an international health consultant in Washington, DC. He relays the question to ProMED, run by the International Society of Infectious Disease, which has over 130,000 subscribers.

February 11 – Provincial Party Secretary Zhang Dejiang said to have reimposed a ban on news coverage, which was violated on February 11 when the *Guangzhou Daily* reported on infections and deaths in the province from atypical pneumonia. On or immediately prior, Politburo Member and Minister of Public Security Zhou Yongkang instructs the PSB on its role in the outbreak. Acting on orders, Guangdong police increase patrols, station officers outside markets to prevent hording of medicines, increase measures against rumors and on-line information that could be "harmful."

Guangzhou remains silent about SARS until a press conference, convened on February 11 by provincial health officials, who report that 305 people have been infected and five have died. (These statistics were later revised to 792 cases and 31 deaths.) They assert the disease is under control, however. During the press conference, Dr. Zhong Nanshan, director of the Guangzhou Institute of Respiratory Disease, names the disease "atypical Pneumonia." When asked if Hong Kong should restrict entry of people coming from Guangdong Province, Huang Qingdao, head of the provincial health department, says, "Atypical pneumonia isn't an unpreventable or untreatable disease. With the right preventive measures, it is absolutely possible to prevent infections. We can see from the measures our province has taken and from the [disease] control situation that we've achieved definite results . . . And up to now, Hong Kong has no reports of illness."

Huang also defends official silence, stating, "Atypical pneumonia isn't a disease we're legally required to report, so we didn't feel it was necessary to make it public. Now, because it has a big social impact, we have decided to make it public." Another report of the February 11 press conference, quotes Huang as saying "it was fine not to tell the public," since epidemics are state secrets.

The first autopsy is performed on a SARS patient at the Nanfang Military Hospital in Guangzhou. Tissue samples are distributed to the Guangdong Center for Disease Control (CDC), Guangzhou CDC, and the Guangzhou No. 8 People's Hospital (which provided the corpse).

February 12 – Nanfang Military Hospital attributes the death of the autopsied patient to a virus-caused pneumonia. The Chinese Academy of Military Medical Sciences (AMMS), which has purview over possible biological and chemical attack, dispatches epidemiologists from the Institute of Microbiology and Epidemiology to Guangzhou to obtain a SARS specimen. The academy sends epidemiologist Cao Wuchun and virologist Zhu Qingyu after obtaining the approval of the General Logistics Department Health Division. Although the PLA No. 1 Hospital, the PLA Guangzhou Military Region Hospital, and the Nanfang Military Hospital all assist, the military researchers could only obtain a thumb-sized lung tissue, some serum samples, and a few drops of saliva.

February 14 – Guangdong Party Secretary, Zhang Dejiang, a Politburo member senior to the Minister of Health, tries to allay public fears when he orders provincial officials to tell the public to "voluntarily uphold social stability, not believe in rumors, and not spread rumors."

Mid-February – Two lung tissues are brought to the Beijing CDC from the Nanfang autopsy. The samples are divided in three. One part is provided to Hong Tao, the CDC's chief virologist and a China Academy of Engineering (CAE) member, who conducts an electron microscope examination. Virologist Li Dexin conducts a polymerase chain reaction (PCR) test. The third part is used for bacterium cultivation. Hong Tao concludes that chlamydia, a common bacterium that is not particularly deadly, is the pathogen for atypical pneumonia.

February 18 - The Chinese CDC holds a press conference to announce Hong Tao's discovery that chlamydia is the pathogen for atypical pneumonia. Some scientists and physicians within the Chinese CDC question Hong's findings and methodology. Doctors in Guangzhou refuse to treat patients with the protocol suggested by the Chinese CDC based on chlamydia.

February 20 – The virologist, Zhu Qingyu, AMMS, Institute of Microbiology and Epidemiology, Beijing, working with colleagues from the Institute of Microbiology and Epidemiology, detects a distinctive halo of spikes, which indicates the coronavirus may be the pathogen for the disease.

Guangdong *Southern Daily*, the official paper of the Guangdong Province Communist Party Committee, reports that provincial health officials realized they had an emergency on February 6 when 45 new cases were recorded on one day. The paper said Guangdong reported the matter to party leadership and the State Council on February 8 where the report was brought to the attention of Wen Jiabao. Wen then dispatched Vice-Minister of Health Ma Xiaowei to Guangdong to investigate.

February 21 – Dr. Liu Jian-Lun, a 64-year-old medical professor at Zhongshan University, who had cared for infected patients at the No. 2 Hospital, Guangzhou, where over 50 medical staff members became infected, travels to Hong Kong. Although he does not feel well, he takes time off to attend a relative's wedding. He and his wife check into the Metropole Hotel in Kowloon on February 21. He stays in Room 911. Johnny Chen and others on the 9th floor of the hotel become infected with SARS, which would spread quickly to different cities and countries.

February 23 – *Washington Post* reports that, following a week of relatively open media reporting, Provincial Party Secretary Zhang Dejiang reimposes the media ban, with the reported support of Hu Jintao, arguing that too much criticism could fuel instability.

26 February – Further testing at the Military Medical Sciences Academy, Beijing, tentatively linked the new coronavirus to the atypical pneumonia, but the chlamydia theory was too well-established to challenge. The findings are not made public.

Late February – Cases in Guangdong Province had doubled from 305 to 792, with 31 people dead. The province did not admit this until March 26, 2003.

Phase Three – SARS Spreads to Beijing, Taiwan, and Mongolia.

March 2003

March 3 – In Hanoi, Johnny Chen, a 47-year old American businessman based in China, became sick. He had arrived from Hong Kong on March 1.

March 4 – Dr. Liu dies in Hong Kong.

March 5 – A 78-year-old woman (Sui-chu Kwan), who had traveled to Hong Kong in February 2003, dies of SARS in Toronto. The Tenth National People's Congress opens in Beijing. Meetings are held during March 5-18.

SARS outbreak begins in Beijing. The first apparent case is a 27-year-old businesswoman who developed symptoms on February 22 while traveling in Guangdong Province. She sought treatment in Shanxi Province, where SARS afterwards developed in two doctors and one nurse who cared for her. After she returns to Beijing, she is hospitalized in a military hospital, and then transferred to an infectious disease hospital. Ten healthcare workers who are exposed to her at the two hospitals develop SARS. Eight family members and friends in Beijing also develop SARS.

March 9 – Vietnam government permits WHO to send additional staff.

March 12 –WHO issues global SARS alert. The announcement comes too late for a WHO employee, Dr. Urbani, becomes infected in Vietnam. The global alert is the first in 10 years, but the alert came "too late" to prevent the spread of SARS around the world.

March 15 – A 72-year-old man who developed SARS symptoms on March 14 while visiting relatives in Hong Kong returns to Beijing on China Air Flight 112. He is evaluated at a Beijing hospital, but not admitted. On March 16, his family takes him to a second Beijing hospital, where he dies on March 20. Fifty-nine cases of SARS infections will be traced to him in Beijing. In addition, China Air Flight 112 is linked to cases in Inner Mongolia and Taiwan. Flight attendant Meng Chunying spreads the infection to her husband, who dies of SARS. She also infects other members of her family in Hohot, Inner Mongolia. Meng considers filing a lawsuit against Air China for withholding information about known SARS exposure, but drops the idea because of a lack of evidence. Among others who were infected on CA flight 11 was Zhu Hong, China Ministry of Trade, who likely infected Pekka Aro, while sitting next to him on Thailand Air Flight 614 from Bangkok on March 23.

March 17-23 - All-Army "Three Defenses" (anti-nuclear, biological, and chemical warfare) rescue training is held in Guangzhou Military Region. Experts from AMMS, Beijing, and other institutes provide the training. (The training did not openly acknowledge any SARS threat, but the biochemical aspects of the emergency training could be applicable to the SARS medical emergency.)

March 19 – Hu Jintao and Wen Jiabao officially become President and Premier, respectively. Zhu Hong falls ill on March 19/20, while in Bangkok, and visits a clinic.

March 20 – Hong Kong health officials link recent global spread of SARS to a guest in the Metropole Hotel. Epidemiologists trace the illness to Professor Liu who was visiting Hong Kong from China. Five other people who come down with SARS also stayed at the Metropole. Some were on the same floor as the professor.

March 21-22 – Following more testing on samples in Beijing, the AMMS, Beijing, is able to develop more evidence to link the coronavirus to the atypical pneumonia cases and report their findings to the General Logistics Department (GLD) and the Ministry of Health (MOH) for approval. Microbial Infectious Disease Institute researcher Zhu Qingyu is credited as the first person in China to isolate the virus from samples taken from victims. His findings are later confirmed in early April by Canadian researchers working in coordination with other scientists.

March 23 – WHO expert team visits Beijing. Zhu Hong travels on Thailand Air Flight 614 to Beijing. He sits in seat 12B, which is next to Pekka Aro in 12A. Aro later told WHO physician Daniel Chin, that Zhu seemed weak and complained of not feeling well.

March 24 – Singapore health minister orders hundreds of people who may have been exposed to SARS into 10-day quarantine. *Stars and Stripes Newspaper* reports that Pacific Command (PACOM) has restricted travel to China and that port calls by the U.S. Seventh Fleet to Southern China and Hong Kong have been cancelled.

March 26 – Zhu Hong is hospitalized in the Ditan Hospital SARS ward. No known attempt is made to contact, screen or quarantine other passengers on the Thailand Air Flight 614, including Pekka Aro, who sat next to Zhu during the flight. Pekka Mykkänen, reporting in the Helsingin Sanomat, quoted an anonymous ILO official who said, "They new that Zhu had SARS. They knew that Pekka Aro sat next to him. But they did not do anything."

Ontario declares a public health emergency and orders thousands of people to quarantine themselves in their homes. There are 27 probable cases of SARS in the province. Toronto begins to bar visitors from hospitals.

The Chinese government acknowledges that the disease has spread outside of Guangdong Province. The news gets low-key treatment, however. Under orders of the city propaganda authorities, the identical three-paragraph story is buried within Beijing newspapers under an optimistic headline reading; "Imported Atypical Pneumonia in Our City Has Been Effectively Controlled." Guangdong officials admit that by the end of February, 15 days after they claimed the disease was under control, cases within Guangdong Province had doubled from 305 to 792, with 31 deaths.

March 27 – Hong Kong quarantines over 1,000 people and closes schools. The Rolling Stones concert in Hong Kong is postponed. Researchers in Hong Kong report they have evidence SARS is coronavirus. They claim to have a quick test for the virus, but Toronto experts question its effectiveness.

Singapore closes its schools. A Taiwanese engineering company closes because five of its employees have SARS symptoms. This causes Taipei to go on medical alert. WHO asks airlines to screen passengers for SARS on flights leaving from Toronto, Hong Kong, Singapore, Hanoi, Taiwan, and Guangdong Province. WHO reports 1,400 cases worldwide, including 53 dead. Ontario health officials order Toronto hospitals closed to visitors, exempting only those who are visiting the critically ill and children.

March 28 – Pekka Aro becomes ill with symptoms of gastroenteritis. He remains in his hotel room, unaware that he is infected with SARS.

Hu Jintao reported to say the Chinese media should do less reporting on official meetings and more on matters that the people care about.

WHO reports 85 new cases of SARS around the world. During a press conference in Beijing, Dr. Henk Bekedam and Professor John Mackenzie, team leader, of a WHO investigation team, say China has basically "become part of [the] SARS global network" and has agreed to provide reports on cases of atypical pneumonia. Dr. Bekedam says, "We are pretty certain that most cases of atypical pneumonia that Chinese authority has recognized from the middle of November until the end of February were indeed cases of SARS." He says, "China has agreed to provide up to date reports of SARS throughout China . . . I would emphasize again that China has agreed to provide updates from all provinces on a regular basis in real time to WHO."

The Chinese government tells WHO it will make SARS a Category B disease, which obligates provincial health officials to notify central health authorities of cases. Although there has been sporadic reporting on the successful handling of SARS, the Chinese media continue to imply that SARS is a distant problem. For example, by highlighting the cancellation of the Rolling Stones concert in Hong Kong because of its serious SARS problems, while downplaying the cancellations of the concert in Shanghai and Beijing, even though it was also because of SARS.

March 30 - The International Ice Hockey Federation (IIHF) announced its decision to cancel the 2003 IIHF Women's World Championship, scheduled to be held in Beijing during April 3-9, citing health risks from SARS.

March 31 – *Wall Street Journal* publishes article entitled, "Quarantine China," which highlights China's initial cover-up and that people going out of China are continuing to carry SARS elsewhere. It called for a travel ban out of Hong Kong and China, and the quarantine of those who have been exposed to SARS. The author wrote: "Isolating a large country would certainly cause economic losses . . . But these have to be weighed against the cost of doing nothing . . . As to panic, information and resolute action are the best antidotes."

Beijing health officials tell a visiting WHO delegation that they have put enhanced SARS surveillance measures in place. In an interview, Hong Tao insists chlamydia is the pathogen for atypical pneumonia, despite evidence to the contrary at the AMMS, Beijing. Chlamydia continues to be the officially authorized theory into April.

The *Beijing Evening News* publishes guidelines on how people can protect themselves from SARS, but provides no context for why this might be necessary. Among the guidelines: maintain good air flow within work and living spaces, avoid crowded areas, wear a 16-layer mask when visiting the sick, wash hands frequently with soap and running water, and seek medical treatment at the first sign of symptoms. The newspaper also advises against randomly taking preventive medicines.

Late March – Long Yongtu, China's former chief trade negotiator, scoffs to a Hong Kong press conference that 300 deaths from SARS (the count at that time) is insignificant for a population of 6 million. He chides the press for being "biased" and causing "anxiety among members of the public.

Phase Four – Cover-up Revealed and Anti-SARS Campaign Initiated.

April 2003

April 1 – The World Economic Forum announced its decision to postpone its annual China Business Summit, scheduled to be held in Beijing on April 14. The meeting was held later in the year during November 6-7.

Pekka Aro seeks medical attention.

U.S. State Department authorizes nonessential personnel and families to leave Guangdong Province. WHO advises travelers to avoid Hong Kong and China. A plane flying from Asia is quarantined in San Jose, CA, after the pilot and several passengers complain of SARS-like symptoms. Emergency vehicles and medical staff garbed in protective clothing meet the plane to examine the passenger. All are later released from the hospital. None are SARS cases.

April 2 – Peka Aro is admitted to the Ditan Hospital.

China reports 361 new cases of SARS for the month of March and a total of 1,153 cases in Guangdong. After some delay, the Chinese permit five WHO experts to visit Guangdong.

News media coverage of Iraq War is reduced, Matt Lauer returned home to NBC, SARS coverage picks up. WHO issues its first travel advisory in its 55-year history, cautioning against travel to Guangdong and Hong Kong. Wu Kejun, Department of International Cooperation of the Ministry of Health, is quoted as telling reporters that "[t]he ministry has required local governments to report to the central government about SARS cases once in a while but how to classify SARS is still under discussion." Shanghai authorities acknowledge a possible SARS case (a cook who had traveled from southern China), but Liu Jun, chief of the Shanghai Health Department, is reported as saying he is unable to recall when the case was identified or which hospital is treating the patient.

April 3 – "SARS Is Nothing to Be Afraid Of" published by Chinese state-run publishing house. Minister of Health, Zhang Wenkang, holds his first press conference on the SARS crisis. He says China is safe, and SARS is under control. He claims there are only 12 cases of SARS in Beijing. Zhang tries to convey the message that it is safe to travel to China. He mocks those who worry about SARS transmissions, saying, "I am confident that all of you sitting here are safe, whether you wear a mask or not." State Council Information Office Vice-Minister Wang Guoqing also criticizes the foreign media for "irresponsible" reporting on SARS that raises fears about the situation in China and Beijing.

April 4 – Jiang Yanyong sends an email to the China Central Television and the Hong Kong-based Phoenix television station accusing Minister of Health Zhang Wenkang of lying. Jiang claims that in the PLA 301 hospital alone, he knows of more than 100 cases of SARS and that six people have died.

Chinese health officials apologize for not being more forthcoming with information. Li Liming, director of the Chinese CDC, says, "We want to apologize to everyone," during a press conference for Hong Kong and Beijing journalists. He says, the failure of mainland state-controlled media to report more fully on SARS has "affected the public's understanding of the illness and their ability to protect themselves." The apology is not covered in the China press.

Health and Human Services Secretary Tommy Thompson and Chinese Minister of Health Zhang Wenkang talk for 45 minutes on the telephone and agree to increase cooperation in the fight against SARS. Japan Ministry of Foreign Affairs issues SARS warning for travel to China, Macau, and Taiwan. Sun Gang, deputy director of China's National Tourism Administration, insists China is safe despite WHO warnings against travel to southern China. Sun claims that tourism during the upcoming May Day Holiday will prove China is safe.

April 5 – There are 163 probable or suspected cases of SARS in the Toronto area, an increase to 149 from the previous day.

April 6 - Pekka Aro dies in Ditan Hospital, becoming the first foreigner to die from SARS in China. At least 24 persons, who were believed to have come in contact with him (UN workers and chauffeurs) are placed in quarantine after his death.

Premier Wen Jiabao meets with the Chinese CDC. The official Xinhua News Agency reports that Wen announces that the Chinese Communist Party and government are making the public's health and welfare their top priority. Wen says government at all levels needs to recognize the complicated and arduous nature of preventing and treating SARS and must be prepared for setbacks. Wen also promises the public health departments will report to the public on SARS at regular intervals.

Residents of Sanlitun Diplomatic compound in Beijing witness a standoff between a man in a car circled by People's Armed Police Hospital staff members who are clad from head to toe in white body suits, trying to prevent the man from leaving his car. The PAP hospital posted a sign a few days earlier announcing the hospital will close for "internal rectification." After several hours, the man is allowed to leave his car and enter a hospital building. His vehicle is driven away. Subsequently, a guard reveals that the hospital has suspected SARS cases. One staff member reports these cases have been taken to another location, but the

PAP hospital remains closed and Chinese authorities release no information to the public.

April 7 – Guo Jiyong of the Beijing Health Bureau was reported to have said that Pekka Aro believed he had contracted SARS during his international flight from Bangkok and that no one who had contact with Aro after his arrival had contracted SARS. Guo said Aro's infection came from outside Beijing. Many foreigners who are able decide to evacuate Beijing.

China claims the outbreak is slowing down, but the number of cases in Hong Kong is climbing—44 new cases are reported, bringing the total to 928. Singapore says it will deploy army medical personnel to help fight SARS and considers installing WebCams in the homes of quarantined persons. David L. Heymann, a WHO official, testifies before the U.S. Congress saying, "We feel that China is taking the measures now they can . . . If these measures had been taken in November, perhaps the disease would not have spread."

Wen Jiabao visits China's Center for Disease Control. The number of suspected and probable cases in Canada reaches 226, of which 188 are in Toronto.

April 9 – Following a leak to *Time Magazine*, the information Jiang Yanyong provided on SARS presence in the PLA 301 hospital is posted on the worldwide web. Various countries in Asia tighten rules on people entering. Malaysia stops issuing entry visas to travelers from China. Indonesia tells its people to stop spitting in public. The Philippines advises against unnecessary travel to Hong Kong or Guangdong Province. Roman Catholic priests in Singapore are asked to stop hearing confession. Taiwan CDC announces that three doctors will travel to Beijing to consult on SARS.

April 9-10 – Non-party experts brief Hu Jintao and Wen Jaibao. Consensus reportedly reached that China should stop covering up and begin working closely with WHO and other agencies.

The number of suspected and probable cases in Toronto rises by 11 to 206. On April 10, Air China flight CA 117 flies from Beijing to Hong Kong with a 71-year-old passenger who is diagnosed with SARS after complaining of illness when she disembarks from the plane.

Noon television report compares the number of Chinese SARS dead (60) to those from traffic accidents (25,395) on Chinese roads during the first quarter of 2003. At the April 10 press conference, Vice-Minister of Health Ma Xiaowei tells reporters that Beijing city has "designated some hospitals with relatively good conditions that are relatively strong technologically, to provide medical services to foreign patients." He says a group of top medical professionals is being assembled to treat foreigners in Beijing.

April 11 – Hu Jintao travels to Guangdong Province. About the same time, Jiang Zemin flees to Shanghai with an entourage that includes Zeng Qinghong and others.

China establishes a formal link to Hong Kong regarding health issues.

Combinations of factors (Iraq War, SARS, terrorist threats, etc.) cause the largest global exodus (1,400 from 17 countries) of U.S. diplomats and families since 1991.

Bi Shengli and Li Dexin, who both were Hong Tao colleagues and oppose his "new variant chlamydia" theory, announce their coronarvirus findings to local newspapers. Both are criticized by Minister of Health Zhang Wenkang for showing disrespect to the official conclusion. They are barred from further publication. Bi Shengli already had been locked out of the Institute of Virology for disagreeing with Hong Tao. Subsequently, the Ministry of Health declares on CCTV that any announcements about SARS that lack the prior approval of the Chinese Ministry of Health SARS Prevention and Treatment Leading Group are unauthorized.

April 12 – Wen Jiabao makes his first visits to a hospital in Beijing that is treating SARS cases. Wen wears no protective clothing and shakes hands with the medical staff. Wen urges the staff to take a "highly responsible" attitude regarding the public's health.

April 13 – Wen Jiabao chairs a national meeting on SARS. He instructs that "China must take resolute measures" to stem the spread of SARS, improve cooperation with WHO and Hong Kong, and keep the world informed on the treatment and prevention of SARS. Wen says it will be "difficult to avoid" SARS having a "temporary impact" on China's tourism, travel, commerce, and international exchanges. He orders that priority must be placed on protecting the health of those attending international events in China.

April 14 – Hong Kong begins screening departing airline passengers for SARS. There are random checks on those entering Hong Kong from China. China announces 4 more deaths, which brings the total to 64. Taiwan Health Minister sends a report on SARS in Taiwan to WHO.

April 15 – Chinese scientists from the AMMS, Beijing, Microbiology and Epidemic Research Institute and the Chinese Academy of Science in Beijing report sequencing the corona virus genome.

Beijing agrees to permit a WHO team to visit Beijing military hospitals.

April 16-19 – Beijing Municipal Government establishes a Joint SARS Leading Group to oversee crisis management through 10 task forces.

WHO reports that two Chinese labs recently joined an international SARS research effort.

April 17 – Hu Jintao calls an unscheduled meeting of the Politburo Standing Committee of the CCP, where he acknowledges that the government has lied and commits the CCP to an all out campaign against SARS. Beijing designates six hospitals for SARS treatment, two of which are military hospitals, which helped integrate military medical care into the overall fight against SARS. WHO reports that military hospitals in Guangdong recently agreed to report their SARS cases, which may set a precedent for other military units.

April 18 – Xinhua News Agency reports a SARS task force has been set up headed by Liu Qi, Beijing Party Secretary and Politburo member. Deputies include Minister of Health Zhang, Beijing Mayor Meng Xuenong, and Deputy Director, PLA, GLD Wang Qian.

April 19 – A ward in Royal Columbian Hospital near Vancouver is closed when a second nurse displays SARS symptoms. British Columbia health care workers now are required to wear goggles in addition to gowns, masks, and gloves. Hong Kong death toll climbs by 12 to 81 deaths. Hong Kong is officially the worst hit location for SARS. The chief executive admits its public health officials were slow to respond to the SARS threat. Apartments, office buildings, food markets, and back alleys are scrubbed. Passengers arriving in Hong Kong's airport must have their temperatures taken. A temperature over 38 degrees celcius becomes a symptom of SARS. The 14th victim of SARS dies in Canada.

April 20 – New Executive Vice-Minister of Health, Gao Qiang, addresses a press conference. He admits to both foreign and domestic reporters that the incidents of SARS are nine times higher than the number reported 5 days earlier (339 versus 37). He adds that Beijing has an additional 405 suspected SARS cases in hospitals. Within 1 hour, Xinhua News Agency releases a two-sentence dispatch stating that Minister of Health Zhang Wenkang (a former military doctor who, reportedly, is a friend of Jiang Zemin) and Beijing Mayor Meng Xuenong both have been removed from their Communist Party posts. Wu Yi takes over as new Minister of Health. Beijing reporters were told both would appear at the press conference that day, but they never appeared.

China reports 12 more deaths and another 400 cases in Beijing—nearly a 10-fold increase. The Chinese government cancels the May Day holiday in an effort to reduce mass movement of people.

Singapore reports a SARS outbreak in its largest vegetable market, spreading fears that the disease will spread into its population of four million. Japan MOFA extends travel warning to Inner Mongolia.

PLA 302 Hospital reported to have discovered that simultaneous basic immunizations and other treatments (hormones, oxygen, anti-viral medicines, and antibiotics) can prevent and treat SARS.

April 22 – The Philippine government institutes screening of U.S. military personnel arriving in the Philippines for the Balikatan 03 exercise. U.S. Air Force (AF) said to have confirmed that two retired AF officers contracted SARS during a trip to Asia in March and recovered, but this was denied by the Air Force on May 5. Experts from the U.S. CDC arrive in Toronto to determine why hospital workers are getting sick despite taking precautions against SARS.

The Mainland Affairs Council (MAC) advocates reducing cross-Strait exchanges because of SARS. Xinhua News Agency releases a speech by Wen Jiabao in which he says that cases of SARS must be reported quickly and accurately, and that "local and departmental leaders will be held strictly responsible" if they do not comply.

April 23 – The Chinese State Council forms a command center for preventing and fighting SARS. A fund of 2 billion yuan is established for fighting SARS.

WHO recommends postponing all nonessential travel to Beijing, Shanxi Province, China, and Toronto, Canada. These locations join Hong Kong and Guangdong Province on the WHO list. One large Beijing hospital with 41 probable cases of SARS is closed. The patients are moved to SARS-designated hospitals. The remaining patients, staff, and visitors are quarantined for 2 weeks.

Construction begins on the Xiaotangshan Hospital[2] in northwest Beijing. Over 1,300 military medical personnel are immediately dispatched from major military regions to work at the facility. During this time, 11 more hospitals will be designated as SARS treatment facilities. Sixty-three hospitals in Beijing also are designated to treat fever patients.

April 24 – A medical emergency command center is established in Beijing. It is organized into a fever clinic that conducts triage of patients and includes designated SARS area within hospitals for specialized care and isolation. Protective equipment is provided to health care workers. Community-based prevention and control measures are established based on detection, isolation, quarantine, and community mobilization. Beijing authorities also have established protocols for triage, isolation, case management and administrative controls, which prohibits visitors to hospitals and separates patients with suspected SARS symptoms from other patients. To address an anticipated shortfall in hospital beds, the 1,000-bed Xiaotangshan hospital is finished in 8 days.

U.S. CDC officials say a travel ban for Toronto is not warranted because public health officials understand the patterns of transmission within the city, but British medical officials support the advisory. SARS forces the closure of a major hospital in Beijing. All public schools in China are also ordered closed for 2 weeks. Another 125 people have come down with SARS in China, and the disease has claimed 110 lives in China.

Taiwan bars people from SARS-affected areas (including China) to enter Taiwan. Chengdu Military Region convenes videoconference on security and stability. Military Region Deputy Commander, Chen Shijun, notes that the security work is more challenging than the previous year. He says that the burden of tasks associated with controlling and preventing SARS are especially burdensome to security forces.

Third-fourth week of April – The SARS outbreak in Beijing reaches its peak. Cases probably number over 100 per day for several days. An increased ratio of patients with no known contact with SARS patients is also reported.

April 25 – Public health officials in Toronto insist that the SARS outbreak is under control. They announce three more people have died, raising the death toll to 19. Ontario health officials say there have been no new probable cases of SARS in the Toronto area since April 9, with the exception of a few hospital workers. Japan announces it will send assistance to China for the fight against SARS. The materials include surgical masks and protective clothing.

April 26 – A 79-year-old woman in Toronto is the 21st SARS victim to die in Canada. WHO says its advisory against nonessential travel to Toronto may be lifted after experts examine new SARS data on April 29.

Although China was reportedly monitoring all passengers on all transport, only cursory, self-monitoring measures are in place at Beijing International Airport on this date. ASEAN health ministers meet in Bangkok, Thailand. During a telephone call, President Bush discusses China's efforts against SARS with President Hu Jintao and U.S.-China cooperation to resolve the North Korea nuclear development issue.

April 27 – *Baltimore Sun* reports that Fort Detrick has been working on SARS since early April 2003.

In Beijing, all patients suspected with SARS had been relocated to designated treatment and isolation areas within hospitals. "At one point, 27 municipalities and 21 district hospitals [are] providing care to SARS patients."

Ottawa announces it will appeal Taiwan's decision to turn back Canadian travelers because of fears they might have SARS. On the same day, Taiwan announces it will temporarily stop issuing visitor and residency visas to people from countries hardest hit by SARS.

April 28 – Beijing government orders residents to stop blocking roads. They had spontaneously blocked people from entering and leaving their neighborhoods and villages out of a fear of the spread of SARS.

"SARS refugees," who began to flee Beijing on or about April 20, continue to leave the city without serious restrictions.

Premier Wen Jiabao attends ASEAN heads of government meeting in Bangkok. This is his first official trip outside China since assuming his position in March. On April 29, ASEAN issued a declaration containing a statement of measures that member countries had committed to in order to share information promptly.

The Chinese State Food and Drug Administration approves clinical testing of a nose spray that was developed by the Academy of Military Medical Science Microbiology and Epidemiology Research Institute to safeguard against SARS. The spray was originally developed for treatment of Hepatitis B and C.

Late April – PLA delegation, headed by General Xing, departs China for official visit to the United States. The delegation members observed quarantine and took western medicines (antibiotics) as a preventive measure prior to their departure.

North Korea suspends its twice-weekly Beijing flights, initiates strict quarantine on land crossings, and stops the ferry from Japan. More than 100,000 visitors are affected.

In Beijing hospitals, daily SARS admissions exceed 100 per day for several days; 2,521 cases of probable SARS have been detected.

April 29 – WHO announces it will lift a travel advisory against Toronto effective April 30. It had been 20 days since a new case and WHO director general, Dr. Gro Harlem Brundtland, says the magnitude of cases in Toronto has decreased. ASEAN-China emergency summit on SARS held in Bangkok, Thailand.

April 30 – Ontario announced two more SARS deaths – a 72-year-old and a 39-year-old, which is the youngest person to die in Canada. Conference on SARS opens in Toronto.

Late April – Chinese officials confidently tell foreign embassy representatives in Beijing that SARS would be finished no later than mid-May, which raises questions about whether or not the Party would attempt to fudge figures to achieve resolution of the problem as quickly as possible.

Phase Five – China's Mobilization Pushes toward "Victory."

May 2003

May 1 – SARS peaks in Hong Kong, Toronto, and Vietnam. WHO officials conclude a SARS conference in Toronto, stressing the need for better international cooperation to control the disease.

Xiaotangshan Hospital officially opens in Beijing. One hundred fifty-six SARS patients from 15 hospitals within the Beijing area are relocated to the hospital.

The PLA Anti-Chemical Warfare Research Academy is praised for producing protective equipment (nose and mouth masks, face masks, protective clothing, boots, and gloves) for over 50 hospitals, as well as Public Security and People's Armed Police units in the Beijing area. The unit also has provided disinfectants to schools and hospitals, and rushed 1,000 sets of special biological protective clothing to the newly opened Xiaotangshan Hospital. The PLA unit additionally has also set up a hotline to provide technical advice on protection measures to relevant units, such as the General Logistics Department, Ministry of Public Health, and Ministry of Public Security.

May 2 – Beijing concurs with WHO officials visiting Taiwan.

Japanese government officials meet to develop an anti-SARS strategy.

May 3 – Two WHO officials arrive in Taiwan to provide assistance.

China reports its No. 361 *Ming* submarine descended on routine training mission killing 70, 2 weeks prior on April 16. The accident was not discovered until 10 days after the accident.

One day after Chinese authorities say the disease has crested, Beijing reports 200 new cases, and nine deaths. Chinese announce they have permitted WHO investigators to visit Taiwan.

May 4 – China says more than one million school children in Beijing will stay home for another 2 weeks. Government officials will conduct classes on television or the Internet.

May 5 – Chinese official television shows Hu Jintao and Jiang Zemin meeting with family members of No. 361 *Ming* submarine, demonstrates unprecedented openness.

May 6 – Authorities in Nanjing order 10,000 people into quarantine as China announces 138 new cases of SARS and 8 deaths. WHO reports that SARS is receding in Vietnam and Canada.

During one of the twice-weekly SARS press conferences in Beijing, Chinese officials read an official statement for 90 minutes and permit no questions. No information is released about what Chinese officials understand how the disease is transmitted, prevention measures, etc.

U.S. Secretary Thompson conducts a phone conversation with the new Chinese Minister of Health, Wu Yi, on how the two countries can cooperate on SARS treatment and prevention. Wu Yi states that Beijing's policy against Taiwan joining WHO has not changed.

The Harbin Polytechnic University reports developing SARS isolation and monitoring cubicles that are self-contained and protected from spreading infection. The General Logistics Department Military Equipment Research Center reports developing improved SARS-resistant protective clothing and masks. The unit has also developed a disinfecting washing machine for bedding and clothing, as well as foodstuffs and drinks for fighting SARS.

May 7 – The Bush administration is reported to have authorized the use of force to detain persons suspected of having SARS, which strengthens the Executive Order signed in April that permits the U.S. Government to quarantine people infected with SARS.

WHO sends a four-person team to two Chinese provinces where it is believed the rural health systems may not be able to cope with spreading SARS.

The first major study of SARS estimates that about 20 percent of the people who are sent to the hospital with SARS in Hong Kong are dying.

The Japanese government directs Health Minister Sakaguchi to establish an anti-SARS system in Japan.

By this time, Beijing's ability to house and treat SARS patients has significantly improved, including the establishment of 63 hospitals for treating fever patients.

May 8 – All probable cases have been concentrated into 16 SARS-designated municipal hospitals. Thirty district hospitals are also providing care for patients with suspected SARS, and more than 60 fever clinics had been established throughout Beijing to triage patients and quickly isolate suspected SARS cases.

WHO estimates that 15 percent of the people who get SARS will die. The rate among the elderly is over 50 percent. WHO issues travel warnings for Taiwan and Tianjin and Inner Mongolia Provinces.

Xinhua reports that 120 officials had been relieved for dereliction of duty.

The U.S. Department of State announces a $500,000 emergency grant to assist China's SARS fight. Secretary Thompson proposes a multi-year, multi-million dollar project to promote collaboration in epidemiological training and to develop greater laboratory capacity in China.

The Japanese Minister of Health sends additional anti-SARS assistance to China.

May 9 – The official number of SARS cases in Beijing is cut from 94 to 48. The reduction eases popular anxiety, and people begin to return to the streets.

In a *Washington Post* commentary, President Chen Shui-bian makes the case that Taiwan should join WHO. A Cross-Strait videoconference on SARS is held. Taiwan press reports U.S. military members have departed Taiwan following the conclusion of the Huanguang Exercise.

May 10 – The PLA Armament Engineering Academy reports developing a thermal imaging infrared thermometer that is capable of taking accurate body temperatures of moving crowds at 30 meters.

May 11 – Japan Minister of Health announces an additional relief package for China.

May 12 – A suspected case of SARS in Finland keeps Canada on the WHO list of countries affected by the disease. The man is Finland's first suspected SARS case and officials claim he got sick in late April while visiting Toronto. Canadian health officials reject this claim.

May 14 – WHO reports on May 13 that SARS has spread to the PLA. Eight percent of Beijing's 2,000 cases—about 150-160 people—are identified as military personnel, but no information is provided on who these people are, or how they contracted the disease.

Canadian Prime Minister Jean Chretien announces that WHO has removed Toronto from its SARS-affected areas list. There have been no new local transmissions for 20 days.

A new fabric for biological protective clothing for medical and health workers fighting SARS is reportedly developed through the joint cooperation of the Shandong Ketele Company, the Academy of Military Medical Science, the PLA Anti-chemical Research Academy, the GLD Blood Products Technical Research Institute, the Beijing University of Chemical Engineering, the Guangdong Microbiological Analysis and Testing Center, and other units. The material has been tested by the PLA Microbiology and Epidemiology Research Institute and other units, which are developing anti-SARS equipment for health workers and patients. The anti-SARS equipment includes clothing that has been developed in cooperation with China Textile Institute. The institute also has developed a new type of positive pressure hood with the assistance of the Academy of Military Medicine Sciences Health Equipment Institute, and an isolation capsule to transport SARS patients.

May 15 – China threatens to execute or impose a life sentence on anyone who breaks SARS quarantine orders or deliberately spreads SARS.

Officers of the China State Anti-SARS Command, report that an anti-SARS positive pressure respiratory protective system that was developed by the General Armament Department Anti-chemical Warfare Research Academy had passed a technical appraisal test and demonstrated that it can filter 99.995 percent of the SARS virus emissions.

May 17 – WHO announces that the SARS epidemic shows signs of ending everywhere except China.

May 18 – The AMMS, Beijing, Hygiene (Health) Equipment Research Institute and Microbe and Epidemic Disease Research Institute report jointly developing an emergency vehicle for handling contagious disease cases, an isolation chamber for transporting contagious patients, a negative pressure ambulance for contagious patients, as well as reusable biological protective clothing and biological protection masks.

The GLD Military Supplies and Equipment (Quartermaster) Research Institute, Anti-chemical Warfare Research Academy, Aerospace Medical Engineering Research Institute, and China Weapons and Equipment Research Academy are praised for developments in protective materials and equipment.

May 19 – WHO issues a travel advisory for Taiwan after it reports a record daily increase in probable SARS cases by 65. The total for Taiwan is 483—the third highest after China and Hong Kong.

May 20 – The total of reported SARS cases at this time is 2,444, with a fatality rate of 6.4 percent.

President Chen Shui-bian calls for a referendum on Taiwan membership in WHO. Beijing blocks Taiwan participation in a World Health Assembly (WHA) panel on SARS.

Chinese military personnel and units are praised for developing new medicines and equipment to fight SARS. Units identified for praise include: AMMS, Beijing; GLD Quartermaster Research Institute; Guangzhou Military Region General Hospital; Guangdong People's Armed Police Hospital; Beijing People's Armed Police Hospital; First Military Medical University; Third Military Medical University, Oral Hospital; and PLA number 301, 302, 309, 320, and 371 hospitals. Personnel who are praised include: Huang Wenjie, Director of the Guangzhou Military Region General Hospital Pulmonary Medicine Department; Zhou Guotai, Deputy Director of the GLD Quartermaster Research Institute; Xu Zhali, Professor of the Fourth Military Medical University, GLD, Xi'an, Oral Hospital; and Zhang Dezhou, Director of Infectious Disease, PLA 371 Hospital.

May 21 – U.S. Secretary Thompson expresses support for Taiwan participation in a WHA panel.

May 22 – The Fourth Military Medical Academy Cell Engineering Center director, Chen Zhinan, reportedly has discovered 9 polypeptides and 13 antibodies that can restrain the coronavirus. The China Center for Disease Control reportedly has tested these and found them to be effective in restraining the SARS virus.

May 23 – Toronto's SARS infection list grows. Canadian heath officials say they are now dealing with at least 25 suspected and probable cases in two Toronto hospitals. Two recent deaths are suspected SARS cases.

China's Association for Relations Across the Taiwan Straits (ARATS) sends a message to Taiwan's Straits Exchange Foundation (SEF) offering aid for SARS fight.

Japan announces closing two plants in China due to SARS.

The Lanzhou Military Region Highland Disease Research Institute reports that it distributed a new book on how to prevent and treat disease in low oxygen environments and disseminated information on "scientific" medicines to its border defense units.

The Anti-chemical Warfare Research Academy is reported to have developed a protective canister for facemasks to protect against SARS. The canister passed the technical appraisal of the State Development and Reform committee, Ministry of Science and Technology and Ministry of Health.

May 24 – At least 500 people in the Toronto area are quarantined as a precaution, while health officials investigate two dozen possible SARS cases. Public health officials confirm they are looking at 33 new cases.

Beijing blocks a Taiwan representative from briefing the United Nations press corps on SARS. The Taiwan Executive Yuan endorses idea of referendum on WHO membership.

The Fourth Military Medical University, GLD, Xi'an, Oral Hospital reports developing, in cooperation with the Xi'an High Oxygen Medical Treatment Equipment Company, two devices that provide oxygen to SARS patients.

May 25 – SEF rejects an offer of SARS aid from ARATS. Two more WHO officials arrive in Taiwan to assist.

May 26 – A *Washington Post* commentary charges that "China's secrecy and dishonesty . . . allowed the SARS virus to become an international problem."

Toronto officials claim the current SARS outbreak has been contained. About 2,200 people are quarantined in Ontario—almost half of these are in Toronto.

Hu Jintao arrives in Russia for an official visit.

The Academy of Military Medical Science reports developing a new anti-viral skin emulsion, which was produced at the Beijing Huitongtianli Biotechnology Company. The product combines biotechnology, nanometric technology, and "disinfecting technology." The emulsion is reported to be the first domestic nanometric-disinfecting product to obtain a State-level health permit.

The Second Military Medical University, GLD, Shanghai, Physiotherapy Research Office reports developing new anti SARS medicines, which were approved for clinical study by the State Food and Drug Control Administration. Clinical trials for a SARS vaccine is reported to have advanced to the animal testing stage in Guangdong.

May 27 – A school in the Toronto area is closed after a student comes down with SARS symptoms. The student has a definite link to the North York General Hospital, the source of the latest SARS outbreak and now closed. The school's 1,500 students and 100 teachers are ordered into quarantine as a precaution.

The *People's Daily* details Beijing's efforts to assist Taiwan with SARS.

May 28 – Two more SARS deaths are announced in Toronto. The Ontario government meanwhile announces that it will spend $720 million to assist healthcare workers and facilities involved in the SARS fight.

A WHA SARS resolution is developed which provides the basis for WHO contacts with Taiwan.

May 29 – The number of SARS cases in Toronto rises as Canada adopts WHO SARS classification method. Doctors in Toronto say the system is simpler and better reflects the extent of the problem. Under WHO's definition, any unexplained case of pneumonia is listed as "probable SARS."

The Third Military Medical University, GLD, Chongqing, reports it has developed a protection filtration face guard for SARS patients. The face guard is said to filter 99.9 percent infected airborne particles. The PLA University has also reportedly developed protective materials for hospital personnel. The PLA Medical Library reports donating computers and Internet equipment to the Xiaotangshan Hospital. The hospital was able to open a long-range medical information mobile workstation at Chanping with access to over 20 databases, 30,000 medical textbooks, and nearly 10,000 periodicals and texts online.

The General Armament Department Anti-chemical Warfare Research Academy reports it developed a protective system for pathological research on SARS. The Academy also has developed a protective system for Ditan Hospital autopsy rooms storing SARS corpses, which the Ministry of Science and Technology and Ministry of Health has approved. The system was developed based on military anti-chemical technology and designed to protect autopsy personnel from infection. The system was development under the urgent initiative of the State 863 Plan to "Research into the Pathologic Anatomy, Specimen Collection, and Pathological Mechanisms of SARS."

May 30 – The Academy of Military Medical Science reports developing a protein chip that detects SARS antibodies. The protein chip can be used to screen and diagnose SARS, as well as for research.

The Chinese Academy of Medical Science and the General Armament Department Anti-chemical Warfare Military Representatives Bureau report jointly developing a "BWT Model Positive Pressure Protective System" that has been operational at the Xiehe Medical Science University since mid-May. The system maintains a zero infection rate among hospital personnel who perform tracheal procedures and medical research, as well as personnel who handle SARS corpses.

May 31 – President Bush signs new legislation regarding Taiwan's admission to WHO.

June 2003

June 1 – At the G-8 conference, Hu Jintao repeats Beijing's opposition to Taiwan independence to President Bush.

June 2 – A review of old cases identifies another SARS death in Toronto, bringing the total to 32. An Ontario nurse calls for an inquiry into how the Canadian health system handled the SARS outbreak.

June 3 – During a Beijing SARS symposium, China's vice minister of health appeals to Asian countries to increase information sharing on SARS. A *New York Times* article dismisses those who believe SARS will do for China what Chernobyl did to USSR (i.e., political change).

June 7 – Ontario health officials announce the death of two more people in the Toronto area on June 6, bringing the total number of deaths in Canada to 33. Canadian officials say 25 of the deaths are connected to the first cluster, which broke out in Canada during March 2004.

June 9 – Japan cancels travel warnings for all areas of China, except Beijing and Guangdong.

June 12 – A consulting firm reports that Toronto's tourist industry has lost nearly $190 million because of the SARS outbreak.

ASEAN heath ministers declare the Asia-Pacific region SARS-free.

June 16 – A total of 190 deaths are reported among 2,053 probable SARS cases in Beijing. The fatality rate is 8.4 percent fatality rate.

June 17-18 – A WHO conference on SARS is held in Malaysia. WHO lifts its travel advisory for Taiwan.

June 18 – Senior Thai and Chinese health officials meeting in Beijing agree to increase cooperation to control of SARS.

The Taiwan CDC director addresses a panel at a WHO SARS conference.

June 19 – In Beijing, a total of 30,172 persons, who had close contact with SARS patients, are quarantined for 2 weeks after their last exposure.

Ontario Province rejects $250 million in SARS relief from the Federal government as insultingly low. The Province seeks Ottawa to cover 90 percent of the estimated $1.5 billion in health-care costs.

June 20 – *Washington Post* reports a crackdown of Chinese media that ended a brief period of relative openness.

June 23 – Ontario announces that two more people have died from SARS. This raises the Canadian death toll to 38 since the outbreak began in mid-March. All are in the Toronto area.

June 24 – WHO removes its travel advisory for Beijing, announcing the situation has greatly improved in the capital since the WHO advisory was issued on April 23. WHO reports that the last new case in Beijing was isolated on May 29, Beijing had been isolated for over 20 days. Cases after this date were ruled out as SARS. Other recent cases were traced to known transmission cases. The report says that for reasons not yet understood, areas of Mainland China experienced a lower fatality ratio than most other outbreak areas, but China's statistical reporting could have skewed the ratio.

June 30 – Canada's deaths from SARS rise to 39 when a 51-year-old nurse, who worked at the North York General Hospital, becomes Ontario's first health worker to die from SARS.

June 29 – Report of Jiang Zemin's invitation for a private meeting in Beijing with former Health Minister is widely interpreted as a sign of political conflict between Hu Jintao and Jiang.

Through end of June – a total of 2,521 patients with probable SARS are hospitalized in Beijing.

July 2003

July 2 – WHO removes Toronto from its list of SARS-affected cities after 20 days have passed since the last known infection. This is double the normal length of incubation for SARS. It is the second time Toronto is removed from the WHO list. Toronto was removed on May 14, but suffered a second outbreak on May 16.

Taiwan is the only remaining country where the disease is still not under control.

July 4 – A total of 8,439 probable cases and 812 deaths from SARS had been identified in 30 countries.

July 5 – WHO Director-General declares that the SARS epidemic is over "for the time being." She says: "We do not mark the end of SARS today, but we observe a milestone: The global SARS outbreak has been contained." Speaking in Geneva, she adds: "This is not the time to relax our vigilance. The world must remain on high alert."

July 11 – The Guangzhou Military Region General Hospital reports it has developed the "BG-95 Nitric Oxide Treatment (breathing) Apparatus" for the treatment of SARS. BG-95 won a State invention patent and an Army Second Class Science and Technology Progress award. The apparatus was approved by the State for clinical application to treat people with respiratory problems. The apparatus is said to have been used successfully during the SARS epidemic for rescuing five seriously ill patients in the China-Japan Friendship and Xuanwu Hospitals in Beijing.

July 21 – An international team of scientists announces it has conclusively identified a corona virus (CoV) as the responsible agent for SARS.

July 26 – The Chinese Minister of Science and Technology, Xu Guanhua, says that although SARS has been effectively controlled around the world, many problems still have not been resolved. SARS still poses a threat. Xu notes that many problems, such as the origin and means of SARS transmission, have not been solved. Xu stresses three key issues that need to be addressed: (1) to clearly determine the source of SARS and its laws of transmission, to provide the scientific foundation and methods to cut off the route of transmission and control the disease; (2) to start research and develop specific, flexible, fast and accurate early diagnostic technology and drug testing, and provide reliable technological methods for virus detection and clinical diagnosis; (3) to accelerate the research and development of efficient medicines and vaccines, in order to effectively prevent and treat SARS.

After the Victory—What Next?

September 2003

September 29 – Ontario's SARS inquiry opens first of 3 days of public hearings in Toronto.

October 2003

October 7 – Dr. David Naylor, dean of medicine at the University of Toronto, releases a report investigating what went wrong during Toronto's SARS crisis. The report, commissioned by Health Canada, suggests Canada needs a public health agency similar to the U.S. Center for Disease Control. It calls for $700 million in new health spending.

2004

January 2004

January 1 – Beijing reports its fiscal revenues are up 18.2 percent for the 9th consecutive year, despite the SARS outbreak.

January 2 – Xinhua reports initial gene sequencing tests show a man with suspected SARS has possible corona virus, according to the Guangdong Provincial center for disease prevention and control. Xinhua reports that during the 2003 outbreak, 5,327 were infected, and 349 died in China.

January 5 – China reports the first case of SARS since the global epidemic was declared over in July 2003. The patient is a 32-year-old television producer working in southern Guangdong Province.

January 7 – A crackdown on *Southern Metropolis Daily*, the first newspaper to report on newest outbreak during late December, is reported:

> The World Health Organization has praised China's cooperation in dealing with SARS since the latest outbreak. But journalists in Guangdong and around the country say that propaganda officials are strictly limiting coverage of the disease to official statements and strongly discouraging the news media from reporting widely on the topic.

April 22-29 – The Chinese government reports a total of nine SARS cases to WHO. Four of these cases are confirmed. All the cases are believed to be traced to the National Institute of Virology, Beijing. (On May 6, after a full investigation of these cases, this was not substantiated since the workers at the lab did not work with SARS samples.) About 1,000 people in Beijing and Anhui Province, the home of one of the victims, are quarantined. WHO praises China for its quick reaction.

May 4 – Beijing confirms three more SARS cases, which confirms all nine cases identified in April 2004 are SARS.

May 18 – WHO declares China's latest SARS outbreak is over after 3 weeks passed without any new infection. The origin of the outbreak remains a mystery, although WHO expresses concerns about biosafety. Of the nine person infected, one died on April 19, 2004, and the others were released from hospital by May 12.

ENDNOTES - APPENDIX I

1. This timeline draws from multiple Chinese and Western sources. The primary sources include: The Center for Infectious Disease Research and Policy (CIDRAP), Academic Health Center, University of Minnesota, *www.cidrap.umn. edu*, accessed December 26, 2004; Ming-Dong and Ann Margaret Jolly, "Changing virulence of the SARS virus: the epidemiological evidence," *Bulletin of World Health Organization*, Vol. 82, No. 7, July 2004, available online at *www.who.int/entity/bulletin*, accessed December 27, 2004; the reporting of Susan V. Lawrence, David Lague, Peter Wonacott, Matt Pottinger, and David Murphy for *The Far Eastern Economic Review* between April-May 2003; John Pomfret's reporting for the *Washington Post* during May 2003; CBC News, "In-depth SARS," *Timeline*, available online from *http://www.cbc.ca/news/background/sars/timeline.html*, accessed February 17, 2004; John Wong and Zheng Yongnian, eds., *The SARS Epidemic—Challenges to China's Crisis Management*, Singapore: World Scientific Publishing Co. Pte. Ltd, 2004; Maryn McKenna, *Beating Back the Devil—On the Front Line with the Disease Detectives of the Epidemic Intelligence Service*, New York: Free Press, 2004; Liang, Zhu, Guo, Liu, He, Zhou, *et al.*, "Severe Acute Respiratory Syndrome, Beijing, 2003," *Emerging Infectious Diseases*, Vol. 10, No. 1, p. 1, available online from *http://www.cdc.gov/ncidod/EID/vol10no03-o553/htm*, accessed September 16, 2004; "WHO Update 83—One hundred days into the outbreak," June 18, 2003, 2004; and WHO Update 87, "World Health Changes Last Remaining Travel Recommendation for Beijing, China," June 24, 2003, available online from *http://www.who.int/csr/don/2003_06_18/en/print.html*, accessed February 17; "WHO Global Conference on Severe Acute Respiratory Syndrome—Where Do We Go from Here?" Summary report of 17-18 June conference, available online at *http://www.who.int/csr/sars/conference/june_2003/materials/report/en*, accessed February 17, 2004, Chinese military and civilian newspaper reporting during January to July 2003 and personal interviews.

2. Xiaotangshan, which was designated as a "field hospital" was built in 6 days and 7 nights using over 7,000 construction workers. Construction took place over an existing hot springs facility at Chanping in northwest Beijing. See Xinhua News Agency report, "Xiaotangshan Hospital Meets Anti-epidemic Sanitary Criteria," published May 2, 2003, and "Xiaotangshan Hospital—Noah's Ark," September 17, 2003, both available online at *http://www.china.org.cn*, accessed January 30, 2005. The 2004 *White Paper* states that a total of 18 military hospitals provided medical treatment to 420 SARS patients. In Beijing, the Xiaotangshan Hospital worked for over 50 days to provide care to 680 patients. It mobilized a total of 1,383 military medical personnel from various military units. See "China's National Defense in 2004," available online at *http://english.people.com.cn/whitepaper/defense2004/defense2004(8).html*, accessed January 3, 2005.

APPENDIX II

PLA UNITS, PERSONALITIES, ACCOMPLISHMENTS, AND COLLABORATION IDENTIFIED IN OPEN PRESS REPORTS ON SARS BETWEEN APRIL-JULY 2003[1]

General Logistics Department—coordinated with Ministry of Health.[2]

- Personalities: Wang Qian, Major General, Deputy Director, GLD.
  - Deputy, SARS task force headed by Li Qi, Beijing Party Secretary and Politburo Member. Included Minister of Health Zhang Wenkang, and Beijing Mayor Meng Xuenong.
- Subordinate Organizations:
  - Health Department.
  - Military Supplies and Equipment (Quartermaster) Research Institute – Zhou Guotai, Deputy Director.
  - Blood Products Technical Research Institute.
    - Collaborated with Shandong Ketele Company, PLA Anti-chemical Research Academy, Academy of Military Medical Sciences, Beijing University of Chemical Engineering, Guangdong Microbiology and Epidemiology Research Institute, among other units, to develop new fabric for protective clothing (reported May 14, 2003).
  - Academy of Military Medical Sciences (AMMS), Beijing.
    - Developed a new anti-viral skin emulsion, which was produced at the Beijing Huitongtianli Biotechnology Company (reported May 26, 2003).
    - Collaborated with Shandong Ketele Company, PLA Anti-chemical Research Academy, GLD Blood Products Technical Research Institute, Beijing University of Chemical Engineering, Guangdong Microbiology and Epidemiology Research Institute, among other units, to develop new fabric for protective clothing (reported May 14, 2003).
    - Personalities. Cao Wuchun, epidemiologist, and virologist, Zhu Qingyu.

    Subordinate units:
    - Health Equipment Institute (Hygiene Equipment Institute).
      - Developed a new type of positive pressure hood in cooperation with Microbiology and Epidemiology Research Institute.

- Institute of Microbiology and Epidemiology.
    - Virologist Zhu Qingyu detected a distinctive halo of spikes on 20 Feb 2003, which indicated the coronavirus may be the pathogen for the disease. Zhu is credited as the first person in China to isolate the virus from samples taken from victims.
    - Adapted Hepatitis B and C nasal spray to SARS – approved by State Food and Drug Administration.
    - Tested new biological fabric developed by collaboration between Shangdong Kelete Company, Academy of Military Medical Sciences, et al. Developed clothing in cooperation with China Textile Institute.
    - Developed anti-SARS equipment for health workers and patients.
    - Developed a new type of positive pressure hood in cooperation with Health Equipment Institute.
- First Military Medical University, GLD, Guangzhou, Guangdong Province.
- Second Military Medical University, GLD, Shanghai.
  - Physiotherapy Research Office.
  - Developed anti-SARS medicines that were approved for clinical study by the State Food and Drug Administration. Animal testing was conducted in Guangdong Province (reported May 26, 2003).
- Third Military Medical University, GLD, Chongqing, Sichuan Province.
  - Oral Hospital.
  - Developed filtration face guard for SARS patients that filters 99.9% of infected airborne particles, and protective materials for hospital staff (reported May 29, 2003).
- Fourth Military Medical University, GLD, Xi'an, Shaanxi Province.
  - Oral Hospital – Professor Xu Zhali.
  - Developed devices to provide oxygen to SARS patients in collaboration with the Xi'an High Oxygen Medical Treatment Equipment Biotechnology Company (reported May 24, 2003).
  - Cell Engineering Center – Chen Zhinan, director.
- PLA Number One Hospital.

- o PLA 301 Hospital.
  - ▪ Doctor Jiang Yanyong revealed the number of known SARS cases at 301 Hospital on April 4 and 9, 2003.
- o PLA 302 Hospital.
  - ▪ Discovered simultaneous basic immunizations and other treatments (hormones, oxygen, anti-viral medicines and antibiotics) can treat and prevent SARS (reported April 20, 2003).
- o PLA 309 Hospital.
- o PLA 320 Hospital.
- o PLA 371 Hospital.
  - ▪ Infectious Disease (Department) – Zhang Dezhou, director.
- o Nanfang Hospital.
  - ▪ First autopsy of a SARS victim performed here on February 12, 2003.

General Armament Department.

- ▪ Subordinate Organizations:
  - o PLA Anti-chemical Warfare Research Academy.
    - ▪ Produced protective equipment for over 50 local hospitals, Public Security and PAP in Beijing; provided over 1,000 sets of protective clothing to Xiaotangshan Hospital; set up hot line to provide technical advice to GLD, Ministry of Health, and Ministry of Public Security.
    - ▪ Collaborated with Shandong Ketele Company, Blood Products Technical Research Institute, GLD, Academy of Military Medical Sciences, Beijing University of Chemical Engineering, Guangdong Microbiology and Epidemiology Research Institute, among other units, to develop new fabric for protective clothing (reported May 14, 2003).
    - ▪ Developed a protective system, which was approved by the Ministry of Science and Technology and the Ministry of Health, for the Ditan Hospital autopsy rooms, which stored SARS remains. The system was based on military anti-chemical technology (reported May 29, 2003).
  - o Anti-chemical Warfare Military Representatives Bureau.
    - ▪ Developed a "BWT Model Positive Pressure Protective System" in collaboration with the Military Medical Science

Academy. The system was in operation at the Xiehe Medical Science University from mid-May (reported May 30, 2003).

- o   Aeronautic (Aerospace) Medical Engineering Research Institute.

Other:

- PLA Medical Library.

    - o   Donated computers and internet equipment for new Xiaotangshan Hospital, Chanping, Beijing (reported May 29, 2003).

- PLA Armament Engineering Academy.

    - o   Developed a thermal imaging infrared thermometer capable of accurately reading moving crowds at a distance of 30 meters (reported May 10, 2003).

- China Weapons and Equipment Research Academy.

- Guangzhou Military Region.

    - o   GMR General Hospital.

        - Developed "BG-95 Nitric Oxide Treatment [breathing] Apparatus" that won a State patent and an Army Second Class Science and Technology Progress Award. The device, which was approved by the State, was credited with saving patients at the China-Japan Friendship and the Xuanwu Hospitals, Beijing (reported July 11, 2003).

        - Pulmonary Medicine Department – Huang Wenjie, director.

- Lanzhou Military Region.

    - o   Highland Research Disease Research Institute.

- Chengdu Military Region.

    - o   Deputy Commander, Chen Shijun – convened videoconference on security and stability on April 24, 2003.

- People's Armed Police.

    - o   Beijing People's Armed Police Hospital.
    - o   Guangdong People's Armed Police Hospital.

ENDNOTES - APPENDIX II

1.   This Appendix is primarily based on military information extracted from Appendix I.

2. Coordination between the General Logistics Department and the China Center for Disease Control does not appear to have been as close as the coordination with the Ministry of Health, as evidenced by the disagreement over the chlymidia theory, which was advocated by CDC chief virologist Hong Tao.

ENDNOTES - CHAPTER 4

1. The author would like to thank Susan V. Lawrence, Andrew Scobell, Larry Wortzel, and Donald Boose, among others who were in China during the 2002-2003 SARS epidemic and preferred not to be mentioned by name, for their input to the research for this paper, as well as comments on early drafts.

2. Albert Camus, *The Plague*, New York: The Modern Library, 1948, p. 34.

3. Other studies of the SARS epidemic in China have ably discussed numerous factors that contributed to the mishandling of this crisis in China during late 2002 to early 2003. Among these factors, for example, were the disruptive role that elite politics played during a time of leadership transition during the November 2002 16th Party Congress and the March 2003 National People's Congress. Additionally, the highly compartmentalized People's Liberation Army played its part by withholding information on the spread of the disease within military hospitals in Beijing by March 2003. A recent publication, John Wong and Zheng Yongnian, eds., *The SARS Epidemic—Challenges to China's Crisis Management*, Singapore: World Scientific Publishing Co., 2004, comprehensively discusses the SARS crisis from the perspective of China's ailing and increasingly commercialized health care system, its lack of a free press, and other economic, legal, and political factors that contributed to the crisis.

4. Rui-Heng Xu, Jian-Feng He, Meirion R. Evans, Guo-Wen Peng, Hume E. Field, De-Wen Yu, Chin-Kei Lee, Hui-Min Luo, Wei-Sheng Lin, Peng Lin, Linghui Li, Wen-Jia Liang, Jin-Yan Lin, and Alan Schnur, "Epidemiologic Clues to SARS Origin in China," *Emerging Infectious Diseases*, p. 1, Vol. 10, No. 6, June 2004, available online at *http://www.cdc.gov/nicdod/EID/vol10no6/03-0852.htm*, accessed September 16, 2004.

5. Maryn McKenna, *Beating Back the Devil: On The Front Lines of the Epidemic Intelligence Service*, New York: Free Press, 2004, p. 236.

6. World Health Organization, "WHO Issues Global Alert About Cases of Atypical Pneumonia: Cases of Severe Respiratory Illness May Spread to Hospital Staff," Geneva: The Organization, March 12, 2003, available from *http://www.who.int/crs/sars/archives/2003_03_12/en/*, cited in Xu, Rui-Heng, *et al.*, p. 1.

7. PLA activities are included within the Appendix I timeline. Appendix II extracts details of PLA units that were mentioned in open press reports, as well as their accomplishments, collaboration with military and civilian activities and some individuals that were recognized for their contributions.

8. Ming-Dong and Ann Margaret Jolly, "Changing Virulence of the SRAS Virus: The Epidemiological Evidence," *Bulletin of the World Health Organization*, Vol. 82, No. 7, July 2004, available online at *http://www.who.int/entity/bulletin/*, accessed December 27, 2004.

9. *The SARS Epidemic*, p. 88.

10. Chinese researchers did not cooperate with international research until mid-April, when two laboratories linked into the network of international labs that were cooperating on SARS. WHO Update 31—Coronavirus never seen before in humans is the cause of SARS, April 16, 2003, available online at *http://www.who. int.csr/archives*, accessed December 27, 2004.

11. John Pomfret, *Washington Post*, May 13, 2003, p. A14.

12. Unconfirmed rumors of continued underreporting of cases persisted during 2003, even after mid-April when the Chinese government initiated its anti-SARS campaign in Beijing and cooperated with WHO officials. There are at least four reasons why the number of reported cases in Beijing and elsewhere may not be completely accurate: (1) Openness may have been limited to an acceptable number of cases, which inhibited reporting; (2) Strict controls on who could say a case was SARS may have lead to some undiagnosed cases; (3) There was considerable pressure at the local level to overcome SARS as quickly as possible. In late April, Party officials readily declared that SARS would be eliminated no later than mid-May, a deadline that could not be met, but which may have provided a disincentive for reporting of potential SARS cases and encouraged local officials to declare success sooner than the disease may have been achieved; and, (4) Poorly managed and compartmentalized information may have resulted in the loss of information, despite best efforts of health care givers.

13. Liang, *et al.*

14. *Ibid.*

15. *The SARS Epidemic*, p. 168.

16. Pekka Aro, an International Labor Organization (ILO) employee, flew from Finland via Bangkok to attend the China Employment Forum in Beijing. He stayed in Bangkok during March 18-23. On March 23, Aro flew on Thai Airways Flight 614 to Beijing. He sat next to Zhu Hong, an official of the Chinese Ministry of Trade. Zhu had flown from Hong Kong to Beijing on China Air Flight 112 on March 15, according to Helsingin Sanomat investigative reporter, Pekka Mykkänen. This flight carried a 72-year-old super-infector, who spread the infection to stewardess Meng Chunying, among others. Meng spread the infection to her husband, who later died, and other family members in Hohot, Inner Mongolia. Zhu complained of illness while in Bangkok and sought treatment at a Thai clinic before boarding the flight, apparently unaware of the risks. According to Mykkänen, Zhu was admitted to the SARS ward at Ditan Hospital on March 26, 2003, two days before Pekka Aro became ill. The reporter believed Chinese authorities were aware that Zhu was exposed to SARS as early as March 20, but they took no action to inform other members of Thai Air Flight 614. Unaware that the man he sat next to on the flight from Bangkok was infected with SARS, Pekka Aro remained ill in his hotel room until April 1, 2003, according to Mykkänen's reporting, when Aro sought medical treatment. He was admitted to the Ditan Hospital SARS ward on April 2 and died four days later. At an April 7 news conference, Guo Jiyong, Beijing Health Bureau, said Aro had immediately sought treatment on March 28, when he

first developed symptoms, and also said that Aro believed he had been infected while abroad. Guo did not mention anything about Zhu's SARS infection or Zhu's ties to the March 15th CA Flight 112 from Hong Kong to Beijing and the Thai Airway Flight 614 on March 23. See "Chinese Premier: SARS Spread Contained in China (April 7, 2003), issued by the Chinese National Tourist Office, April 7, 2003, available online at *www.ChinaTravelNews.com* [accessed May 31, 2005]; "Announcement of the Death of Pekka Aro," Disabled People's International Current News and Updates, May 3, 2003, available online at: *www.dip.org* (accessed May 31, 2005); and "China withheld vital information from Finnish SARS victim," by Pekka Mykkänen, Helsingen Sanomat, May 27, 2003, available online at *http://www2.helsingensanomat.fi/english/archive/news* (accessed May 31, 2005).

17. Gina Kolata, "Flu: The Story of the Great Influenza Pandemic of 1918 and the Search for the Virus that Caused It"; William H. McNeal, *Plagues and Peoples*, New York: Doubleday, 1977.

18. *The SARS Epidemic*, p. 172.

19. Chen Huai, Director of the Development Research Center of the State Council Research Institute for the Market Economy, in FBIS Compilation Supplemental of PRC Central and Provincial Media Reports on Severe Acute Respiratory Syndrome for July 9-22, 2003. Also see John Wong, Sarah Chan, and Liang Ruobing, "The Impact of SARS on Greater China Economies," *The SARS Epidemic*, pp. 11-43.

20. See "Severe Actute Respiratory Syndrome—Press Briefing, Beijing, China, March 28, 2003, available online from *http://www.who.int/csr/sars/2003_03_28/en/ print.html*, accessed February 17, 2004, for example, which stated that the Chinese had agreed to provide up to date reports on SARS cases for all provinces on a real time basis to WHO.

21. Fred Hiatt, "Lies in the Absence of Liberty," *Washington Post*, April 14, 2003, p. 17.

22. John Pomfret, "Outbreak Gave China's Hu an Opening—President Responded to Pressure Inside and Outside Country on SARS," *Washington Post*, May 13, 2001, p. A14.

23. During the late summer of 2002, these restrictions on party members against talking about possible leadership changes at the upcoming 16th Party Congress were reinforced based on personal direct and indirect discussions by the author with Party members in Beijing.

24. Joseph Fewsmith, "China and the Politics of SARS," *Current History*, September 2003, pp. 250-255. Also see James Mulvenon, "The Crucible of Tragedy: SARS, the Ming 361 Accident, and Chinese Party-Army Relations," *Chinese Leadership Monitor*, No. 8, Fall 2003, available online at *http://www.chinaleadershipmonitor.org*, accessed October 29, 2004. Mulvenon argues that Hu Jintao "appear[ed] unwilling or unable to directly challenge Jiang Zemin's leadership" during the SARS crisis, however, I would stress that Hu skillfully and realistically took advantage of the SARS crisis to promote his power base without causing disruptive cleavages within the Party or even between the Party and Army.

25. Also see Appendix II, which highlights PLA actions beyond mere propaganda efforts.

26. The near-symbiotic relationship between the PLA and CCP rests upon the subordination of the military to the Party, but also depends upon Party penetration throughout the military since military officers and others within the military are Party members. The Party's leading role is also strengthened by the prominent role of the General Political Department, which is represented down to the lowest level by the political officers.

27. "World Health Organization Update 32—Situation in China and Hong Kong, status of diagnostic tests," April 17, 2003, available online from *http://www. who.int/csr/sars/archive/2003*.

28. The *SARS Epidemic*, p. 88-89.

29. The PLA was not the only organization having difficulty obtaining specimens in February. Chinese news reported hospitals in Guangdong Province refused to give specimens to representatives from the Beijing Genomics Institute (BGI), for example, who were forced to steal specimen. (See *The SARS Epidemic*, p. 166.) Among the reasons for this possessiveness was economic competition, since hospitals and research units hoped to patent a "cure" for SARS. There may also have been a shortage of specimens since a limited number of autopsies were conducted worldwide out of a fear of infection. Canada appears to have led the world, with 21 autopsies conducted in Toronto. See Anne-Marie Tobin, "SARS Coronavirus Found in Lungs, Other Organs of Those Who Died: Autopsies," December 20, 2004, reported online at *http://news.yahoo.com*, accessed December 26, 2004. The biggest problem, however, was a lack of leadership that would have directed coordinated investigation of the disease in China.

30. *SARS Commission—The Testimony*, September 29, 2993, accessed July 1, 2004, available online from *http://www.cbc.ca/news/background/sars/sars_commission.html*.

31. Joseph Fewsmith, "China and the Politics of SARS," *Current History*, September 2003, pp. 250-255; John Pomfret, "Outbreak Gave China's Hu an Opening—President Responded to Pressure Inside and Outside Country on SARS," p. A1. Also see James Mulvenon, "The Crucible of Tragedy: SARS, the Ming 361 Accident, and Chinese Party-Army Relations."

32. See Appendix 1, 2004, Addendum to *Timeline*. China's performance during late December 2003 and early 2004 was similar to the original outbreak of SARS, although the timeline for action was shortened. The later outbreak in the spring, however, was more promising, although problems remain in China's ability to handle biohazards, and the source of the infection could not be fully traced.

CHAPTER 5

CHINESE DECISIONMAKING UNDER STRESS:
THE TAIWAN STRAIT, 1995-2004

Richard Bush

Introduction.

Several times during the last decade, tensions in the Taiwan Strait rose to a dangerous degree. In each case, the People's Republic of China (PRC) concluded that actions by Taiwan threatened its fundamental interests and that action, not excluding some kind of military action, was necessary to demonstrate Chinese seriousness and compel a Taiwan retreat. Because the possibility of conflict—at least as the result of accident or miscalculation—was not trivial, the United States intervened to reduce that possibility and restore stability. The episodes, in summary form, are as follows:

- In June of 1995, Taiwan President Lee Teng-hui made a visit to the United States. He spoke at Cornell University, his alma mater, about the island's democratic transformation after decades of authoritarian rule. China irately suspended the semi-official contacts that had developed with the Taiwan government and engaged in military exercises to show its displeasure. It also downgraded its relations with the United States because Washington had allowed Lee to visit in the first place. Then in March 1996, at the time of the Taiwan election, it mounted even more aggressive displays of military force, including the launch of ballistic missiles to targets outside of the island's ports. The United States, concerned that war might occur through accident or miscalculation and that China might misread its own resolve, sent two aircraft carrier battle groups to the Taiwan area.

- In July 1999, Lee Teng-hui announced in a press interview that cross-Strait relations were between two states (or countries—the Chinese term that Lee used is ambiguous).

135

China then unleashed a propaganda barrage against Lee. Chinese fighters patrolled further out over the Taiwan Strait than usual. To prevent tensions from escalating, the United States sent diplomats to both Beijing and Taipei to encourage restraint.

- In March 2000, it became clear that Chen Shui-bian, the candidate of the Democratic Progressive Party (DPP), a party that had advocated establishment of a Republic of Taiwan completely separate from China, might win the election for president. At a press conference a few days before Taiwan voters were to cast their ballots, China Premier Zhu Rongji declared in threatening tones that, "Taiwan independence means war." Because Beijing had already in effect declared that Chen was the candidate of Taiwan independence, Zhu's bluster suggested that a Chen victory would be a *casus belli*. Chen did win the election, and Washington again sent envoys to urge restraint.

- In 2002 and 2003, Chen Shui-bian, as part of his campaign for re-election, made a series of statements that Beijing interpreted as evidence that he was preparing to "break out" of the status quo. It believed that his proposals to institute referenda and write a new constitution signaled that he would make Taiwan independent under the cloak of democracy, and so issued increasingly shrill warnings. The United States sought in various ways to dissuade Chen from this course, culminating in President Bush's criticism of Chen on December 9, 2003. Beijing responded to Chen's subsequent victory more calmly than it did 4 years before, yet the belief became increasingly common in China that military conflict was inevitable.

We may debate whether all of these episodes necessarily meet the Brecher and Wildkenfeld definition of a crisis, that is, a situation that (a) is a threat to basic values or core interests; (b) produces a sense of urgency in devising a response; and (c) has the potential for military conflict, large-scale violence or mass casualties. In all four, Beijing perceived that there was an acceleration of a trend toward the permanent separation of Taiwan ("independence") and foreclosure on the aspiration of national unification. That outcome was utterly unacceptable because the Chinese Communist Party (CCP) regime had bound its legitimacy to "returning Taiwan to the Motherland." Arguably, the first two episodes better qualify, because the PRC responded with displays of force and so they had the potential to lead to conflict, at least through accident or miscalculation. The other two did not include that critical element, yet even they were not trivial because they included Chinese rhetorical threats to use force. The Clinton administration certainly took the 2000 case seriously, and the Bush administration worked hard to contain the 2003-04 one. The degree of urgency is also a question mark. The 1995-96 and 2003-04 episodes occurred over a period of months. Moreover, in no case was there a specific Taiwan action that Beijing required before tensions could decline (analogous to withdrawing missiles from Cuba). Often the objective was to reshape the environment, slow the unwanted trend or reduce its obviousness, and influence the views of the United States.

So these episodes may fall within the mini-crisis or near-crisis part of the spectrum. Yet they all placed the Chinese system under some degree of stress. And the fact that Taiwan is at the center of each allows us to assess PRC behavior over the range of cases. Is there a common Chinese approach to stress or crisis management in all of them or do we observe some degree of learning?[1]

Context: Security Dilemma.

At the outset, it is worth observing that Taiwan and the PRC are trapped in a security dilemma. Each sees the power of the other and fears how it might be used. Each takes steps to guard against that threat, only to trigger a hedging response from the other side. Thus Beijing and Taipei each add new systems to their respective arsenals to counter the acquisitions of the other. In the 1990s, the PRC acquired advanced fighter aircraft from Russia (the Sukhoi-27s and 30s) and Taiwan secured F-16s from the United States and Mirage 2000s from France. Over that same decade, Beijing bought Kilo-class submarines from Russia and Taiwan requested diesel-powered submarines from the United States. The PRC produced indigenously a growing force of short- and medium-range ballistic missiles, and Taiwan sought to acquire missile defense capabilities—and did get Patriot batteries—from the United States. In addition, the Taiwan armed forces worked to improve institutional ties with their American counterparts.

This state of affairs has a long history, of course. Yet the series of episodes that began with Lee Teng-hui's U.S. trip demonstrates that, even though the two sides of the Strait are engaged in something of an arms race, this is not the classic security dilemma as described in the international relations literature. It is not a simple case where Beijing fears that Taipei's arms acquisitions make it more vulnerable to attack. What Beijing dreads instead are Taiwan political initiatives to permanently separate the island from China, or, as they might put it, seizing Chinese national territory by fiat rather than force. Taiwan's military power and its *de facto* alliance with the United States become relevant not because they are inherently threatening but because they are seen as useful in defending those political initiatives. It is at least to deter those steps and to counter Taiwan's defensive military build-up that the PLA acquires new capabilities. And it is supposedly to allay those fears that Beijing has asked Taipei to reaffirm the one-China principle. The central political dimension of this security dilemma gives it an asymmetrical and perhaps unique character. As Thomas Christensen so elegantly puts it:

> Security dilemma theorists have assumed that international security politics concerns merely defending sovereign territory from invasion and foreign acquisition. [But] to a large degree, the Taiwan question is one more of the island's political identity than of the PRC's territorial expansion. The danger to the PRC is that Taiwan might eventually move from de facto independence to legal independence, thus posing an affront to Chinese nationalism and a danger to regime stability in Beijing.[2]

This security dilemma has another political dimension. That is the PRC's use of united-front tactics within Taiwan to try to prevent what it most seeks to avoid (independence) and perhaps secure what it seeks (unification on its terms). Beijing's effort to manipulate Taiwan politics aggravates the anxiety felt by some segments of the Taiwan public and leads them to suspect the worst of PRC intentions. (Note the asymmetrical nature of this situation; Taipei doesn't have the option of meddling in Chinese politics).

An important dimension of this dilemma is a heavy overhang of mistrust. Not only does each side watch the actions of the other and take steps to deter the worst, neither believes that the other will keep its word, no matter what promises it might make.

For Taipei, the dilemma is even more profound. It has relied on the United States for its security since 1950 and will continue to do so because it is in an increasingly weak position militarily. With this quasi-alliance come the problems of any alliance: fears of abandonment and entrapment. Taipei worries, sometimes to a paranoid degree, that Washington will, either by accident or design, sacrifice its interests for the sake of relations with Beijing. Washington worries sometimes that the initiatives of Taiwan's leaders will drag it into an unnecessary conflict with China.

Context: PRC (Mis-)Perceptions of the Taiwan Threat to Its Interests.

If it is Taiwan political developments that drive the security dilemma for China, then how Beijing defines the threat that those developments pose becomes important in assessing its crisis management. This is a complicated issue, but Beijing's working hypothesis has been that Lee Teng-hui's goal since at least the early 1990s, and Chen's since before he became president, was to obstruct unification and permanently separate Taiwan from China. That general perception has colored China's reaction to specific initiatives and events and so contributed to the level of stress it felt in each of the cases under review.

There are, to be sure, people on Taiwan who would like their homeland to be a totally independent country with no special relationship to China, and they exert pressure within the political system. Even more Taiwan people would share that objective if it could be achieved without cost (military action by China). There is no question that Lee Teng-hui and Chen Shui-bian have during their presidencies said things and taken actions that violate Beijing's sense of how they should talk and act and reinforces its own fears of a break-out.

Yet my own research on their statements and actions suggests that Beijing has generally misunderstood Lee's and Chen's agenda (and that of the Democratic Progressive Party, which Chen leads). Contrary to the PRC view, Lee Teng-hui did not, in fact, oppose unification in principle. Instead he took a firm and consistent stand on the terms and conditions that would define Taiwan and its unification with the mainland. And although Chen led a traditionally pro-independence party, his formal statements clearly preserve the option of certain kinds of unification. For both presidents, the question was not so much whether Taiwan was a part of China but how it was a part of China. For them, one country, two systems was simply not an acceptable answer to that question, because it would place the post-unification Taiwan government in a subordinate status vis-à-vis the central government in Beijing. In their view, their government is a sovereign entity, and the PRC has to take that into account. Beijing, however, has interpreted Taiwan initiatives that are contrary to one country, two systems as separatist. It tends to reject the idea that national unification can occur among sovereign entities (which it has). Of course, designing a workable confederation, federation, or commonwealth for the cross-Strait context would not be easy. But that is different from saying that Lee and Chen were *ipso facto* separatists because they adopted the substantive approach that they did. A less narrow approach to the question of national unions might yield a different outcome. A different Chinese approach to *how* Taiwan was part of China might produce Taiwan acceptance that it *should be* a part of China.

Some might argue that Lee and Chen said one thing and did another. Thus Beijing has cited an array of their actions as further evidence of separatist intent (such as, reform of the political structure, fostering a Taiwan identity, purchasing weapons from the United States, and seeking more international space). Yet most of these steps were important for their own sake, and responded to demands from within Taiwan society. The purchase of arms from the United States was, in part, a response to Beijing's own military modernization and its refusal to renounce the use of force. All were consistent with Taipei's definition of its status as a sovereign government. None constituted *prima facie* proof that Lee and Chen intended to separate Taiwan from China permanently—unless the formula for unification excluded those elements, as China's did. If one looks at specifics, moreover, Lee's actions—his 1995 visit to the United States, for example—do not necessarily have the meaning that Beijing and others attribute to them.

Moreover, Lee and Chen were right in the mainstream of Taiwan views since democratization on the basics of cross-Strait relations. Public opinion and all major political parties shared their view that the government possessed sovereignty, that the people of the island had a right to be represented in the international system, and that the PRC's growing military capabilities were an obstacle to reconciliation. Lee and Chen helped shape that opinion, of course. They exploited the fears of Taiwan people toward China's threat to the island's security and political well-being. But the consensus they fostered would not have been possible had such sentiments not already existed in latent form and Beijing not taken steps designed to intimidate the Taiwan populace.

Beijing's definition of Lee and Chen as separatists became the lens through which it interpreted any new actions, particularly the unexpected and the departure from routine practice. Those initiatives became evidence of the acceleration of a negative trend, one that would only get worse if it was not nipped in the bud. Lee's American trip, his "two-state" formulation, Chen's 2000 election, and his re-election proposals all signaled, in PRC eyes, a looming break-out that required rapid and decisive action if the situation was to be contained.[3]

As an aside, it is worth noting that the Taiwan initiatives that Beijing found so provocative did not occur in a vacuum. Both Lee and Chen, each in his own way, were responding to a rigid PRC stance that continued even after they, in their minds, sought to demonstrate an openness to some form of unification. Each responded to Chinese recalcitrance with mounting frustration and toughening of their respective approaches. Thus, Lee Teng-hui's decision to pursue a visit to the United States was driven in part by his unhappiness with Beijing's strong opposition to his search for a flexible reentry into the international system. There is evidence that he made his state-to-state pronouncement in part because he had received information that Beijing would make an announcement that would frame cross-Strait relations in a way that would put Taiwan on the defensive. He therefore decided that he should preempt that statement with one of his own.[4] One can make a persuasive argument that Chen's initiatives during the 2003-04 election campaign reflected not only his compelling need to rally his political base but also his frustration that Beijing had not reciprocated his effort to steer a middle course during the previous campaign and the first 2 years of his presidency. Lee's and Chen's reactions to Beijing's actions then fostered an even harsher Chinese response.

To sum up the argument so far: China and Taiwan are locked in a security dilemma in which Beijing fears political initiatives by Taipei that would foreclose its goal of national unification and permanently separate Taiwan from China. For at least the last decade, it has had a strong bias to believing that Taiwan's leaders have just such a separatist intention, even though there is another explanation for their behavior. And Beijing ignores the possibility that its own actions may be stimulating that behavior and the resulting spiral of hostility. How do these contextual factors condition China's response when a stress-inducing situation occurs? The decisionmaking system, an absence of cross-Strait dialogue, domestic politics, and the role of the United States are all relevant factors.

The PRC Decisionmaking System.

Michael Swaine offers the most thorough description of PRC policymaking concerning Taiwan. He concludes that during routine periods the policy process on Taiwan has become "highly regularized, bureaucratic, and consensus oriented." As the issue has become more complex and its salience for senior leaders increased, so too has the number of actors and their responsibilities. Decisionmaking is characterized by "extensive horizontal and vertical consultation, deliberation, and coordination." At the highest levels, discordant views are harmonized or muted through a process of informal deliberation among the senior leadership.[5]

Yet Swaine's information suggests a policymaking process that remains fairly centralized. Jiang Zemin dominated Taiwan policy during his time as state chairman and did so until he gave up the chairmanship of the Central Military Commission in September 2004. The consensus norms of the Politburo Standing Committee (PSC) created some constraint on him, but not an absolute one. Line agencies provide information and carry out instructions but appear to have no role in policy initiative or formulation. This is consistent with Chinese foreign policymaking as a whole, which Ning Lu describes as "highly centralized and . . . very much personalized.[6] Moreover, Swaine suggests that in a crisis, centralization is even more pronounced. Then the PSC asserts itself relative to the Taiwan Affairs Leaders' Small Group, and senior military leaders and the Central Military Committee (CMC) participates in decisionmaking that has a military dimension.[7]

The danger of this centralized, personalized policy process that already starts with a general misunderstanding of Taiwan intentions is that senior-level officials of the foreign-policy process will "highjack" the policy response. Because those leaders are senior, they are unlikely to be challenged by lower levels whose understanding is more nuanced, but instead will receive deference and obedience from their subordinates. To be sure, this phenomenon is not unique to the PRC. It is common with all actors, including Taiwan and the United States. But it is a tendency to which China is particularly prone.

We can see the dysfunctional character of the PRC decisionmaking system in how it performed in the Taiwan Strait crisis of 1995-96. The consensus of a number of scholars on that episode is as follows:[8]

- The PRC leadership believed that Lee's actions to expand Taiwan's international space reflected his intention of "Taiwan independence."

- It regarded Lee's activities and American complicity in them as a threat to China's vital interests.

- It chose to employ coercive diplomacy to demonstrate China's serious resolve, to compel Taiwan and the United States to reverse course, and to deter other countries from following the U.S. lead.

- It concluded after the fact that the benefits of the action outweighed the liabilities.

In evaluating the quality of Chinese decisionmaking during this episode, there are several questions that must be addressed. First of all, was this simply a case of a PRC reaction to a Lee Teng-hui action? Although it is a staple of PRC foreign policy rhetoric that it is always others (not China) that create problems, I have suggested that the Taiwan actions to which the PRC reacted may, in fact, have been Lee Teng-hui's and Chen Shui-bian's *response* to Beijing's rigid approach to cross-Strait relations (concerning unification formulas, Taiwan's international space, etc.).

Second, did Beijing accurately perceive Lee Teng-hui's intentions when it concluded that he was pursuing an independent Taiwan? By and large, the scholars cited here tend to accept the PRC's assessment. Yet as discussed above, Beijing's narrow approach to unification (one country, two systems) ignored approaches to unification that were more Taiwan-friendly and fostered a very expansive definition of what constituted independence.

Third, did the PRC accurately judge U.S. intentions? It appears that Beijing believed what it wanted to believe: that Washington would not allow Lee Teng-hui to visit the United States. But Warren Christopher clearly tried in late April 1995 to signal Vice Premier Qian Qichen that the Congress might well take the decision out of the Administration's hands. Washington's dual signals may have continued, but China can be faulted for not taking seriously the negative warnings. It is also worth noting that if Lee was not pursuing a secessionist agenda, then the United States could not have been supporting such an agenda either, as Beijing claimed. Indeed, the Clinton administration was the victim of domestic political pressure to treat Lee well, pressure that Lee himself had stimulated. A more controversial question concerns the PLA's summer 1995 exercises, which it conducted to show China's displeasure and resolve concerning Lee's trip. Did Beijing interpret the relatively mild American rhetorical response as an invitation to act more aggressively later on? On this question, on which there may be no early answer, there is obvious disagreement. It does seem, however, that the dispatch of the two carrier groups in March 1996 caught Beijing by surprise.

Fourth, did Lee Teng-hui's course of action threaten China's vital interests? If Beijing misperceived and exaggerated Lee's intentions, then its assessment of China's interests was probably flawed. If Lee truly sought the permanent separation of Taiwan from China, then his success would thwart China's goal of national unification—or at least undermine the legitimacy of the CCP, as it defined it. But if his goal was to not to reject unification *per se* but to question Beijing's terms and conditions, then Beijing miscalculated the challenge he posed.

Fifth, was coercive diplomacy the appropriate way to deal with the challenge that Lee represented? Again, if the definition of the problem and assessment of the affected interests were flawed, then the action taken could well be flawed also. In a way, PRC leaders were attacking symptoms of the problem and not addressing the cause of the problem itself. They responded to Lee Teng-hui's high-profile travel (and American cooperation) and succeeded in discouraging subsequent trips. But if the underlying problem was a rather fundamental difference of opinion on *how* Taiwan might be a part of China, then perhaps coercive diplomacy only exacerbated the problem.

Finally, how should we assess the balance of costs and benefits of its displays of force? Chinese spokesmen admit there were costs. Beijing has been more cautious in subsequent elections about taking even low-level military steps because it has gradually realized that its actions in 1996 (and statements in 2000) were counterproductive, in that they probably strengthened the political positions of those they opposed. On the other hand, the overall Chinese assessment is that benefits outweighed costs. Robert Ross reports that this was the conclusion of the leadership.[9] Yet if we look again at the two sides of the ledger above and do so more objectively, it is difficult to avoid the conclusion that on balance the negatives outweighed the positives. Lee Teng-hui was more restrained, and the United States reaffirmed the basic principles of its China policy and became more engaged in encouraging Lee's restraint. Taiwan understood better the dangers of crossing China's bottom line, but it was also more hostile to the idea of unification. Other countries, including the United States, were less likely to cooperate with Lee Teng-hui's international initiatives, but they were more worried about Chinese intentions and the prospects for conflict. (That Beijing sees the episode as a relative success should itself be a cause for concern.)

We see the same rush to judgment and, sometimes, rush to action in other episodes. In 1999, Beijing interpreted Lee Teng-hui's state-to-state formula as a new step toward permanent separation and undertook more aggressive PLA Air Force (PLAAF) patrols in the Strait. Yet a case can be made that Lee was merely making explicit what was implicit in Taiwan's long-standing position in anticipation of political negotiations. How he deployed that formula was inappropriate (and his staff knew it), but the content was not really new. In early 2000, Beijing suddenly woke up to the possibility that Chen might win and responded by threatening war (Zhu Rongji may have thus secured Chen's victory). During late 2003 and early 2004, Beijing took an absolutist approach to constitutional revision (in contrast to the United States, which focused more on problems of substance and process).

It is true that as a result of the 1995-96 crisis and the 1999 and 2001 episodes, the PRC has sought to remedy the defects in its decisionmaking process concerning Taiwan. For example, the agencies responsible for interpreting developments on the island are probably more accurate in their analysis of events and their significance. The temptation to over-react has been resisted. Beijing showed greater restraint as the campaign for the March 2004 presidential election unfolded.[10] Yet Robert Suettinger's judgment on how the new Chinese leadership would cope in a future crisis is probably on the mark. "At some point, . . . Hu [Jintao] and Wen [Jiabao] may find themselves in a situation in which they need reliable information, short time-frame decisions, and sound judgment on a foreign policy issue. It is fair to wonder whether the decisionmaking system currently in place in China—opaque, noncommunicative, distrustful, rigidly bureaucratic, inclined to deliver what they think the leaders want to hear, strategically dogmatic, yet susceptible to political manipulation for personal gain—will be up to the task of giving good advice."[11]

The Politics of National Security.

Chinese decisionmaking, on Taiwan and anything else, takes place within a political context in which leaders must take account of both the views of their colleagues and competitors and the public mood. By virtue of history, nationalism, and foreign policy, therefore, a PRC leader who is perceived to have mismanaged the Taiwan Strait issue and to have been duped by the United States is a leader who is vulnerable to criticism from his colleagues, from foreign policy experts, and from the public. The danger for the party is that its already weak legitimacy would be further undermined. Regaining the island is the brass ring of Chinese politics; to somehow "lose" Taiwan can be the kiss of death. How the leaders respond to these political forces can affect crisis-management.

On the elite level, Jiang Zemin was fairly free to call the tune concerning Taiwan from around 1994, and did so until he stepped down from his various positions, beginning in the fall of 2000 and departing the final one in March 2005. The only major exception to his dominance came in 1995 when some of his colleagues criticized him for allowing the Lee Teng-hui visit to the United States. Even in this case, information is limited and Western scholars disagree to some extent on the degree of elite conflict. Robert Suettinger concludes that during the 1995-96 crisis "there can be little doubt that leadership frictions and competition for power continued throughout the period, and may have intensified, given the high tension of the situation." At least two of Jiang's civilian political rivals—Qiao Shi and Li Ruihuan—used the Taiwan crisis to put Jiang on the defensive.

What is less clear is whether there was a split between the military and the rest of the leadership on how to respond to the Lee visit. One school of thought, represented by Suettinger, John Garver, and Tai-ming Cheung, concludes that the military had opposed civilian policies for some time and used events like the Lee visit to impose their views on Jiang, constraining his options and forcing a tougher policy that employed training exercises as tools for intimidation. They, along with some civilians, constrained Jiang's options. Others, particularly Michael Swaine and You Ji, tend to dismiss the idea of a deep division over Taiwan. They see a consultative policy process (not a factional one), in which the leadership altered its policy consensus to respond to changing circumstances. Both civilian and military leaders agreed that a tough response to Lee's visit was required. The military was one participant in that process and had a relatively significant impact when national security issues were on the agenda. Actors differed on the timing and nature of the response. Civilians stressed diplomatic and political measures, while military officers favored military ones. In a more recent assessment, Jing Huang says there are differences between civilian and military leaders but attributes them to institutional differences. Civilians focus on containing crises; generals prefer to show strength and resolve (which could exacerbate the crisis).[12]

In the summer of 2004, there were reports of policy conflict between Jiang on the one side, and Hu Jintao and Wen Jiabao on the other.[13] Most observers concluded, however, that the toughening of the Chinese position was primarily a response to what the Chinese regarded as a deteriorating situation and secondarily an effort by Jiang to reaffirm his authority as CMC chairman. There was no observable difference between him and Hu and Wen on the substance of Taiwan policy.

Beyond the elite, Jiang appears to have developed a skillful approach for responding to outpourings of nationalistic sentiment over perceived external challenges on controversial issues like Taiwan. The most sophisticated case study of this pattern remains David Finklestein's analysis of the "peace and development" debate of 1999, which occurred in the aftermath of the accidental North Atlantic Treaty Organization (NATO) bombing of the Chinese embassy in Yugoslavia, when righteous indignation and moral grievance combined to fuel the most serious demonstrations since Tiananmen.[14] Jiang did not try right away to suppress the public response and did not restrict discussion on call-in shows, thus allowing the public to vent its anger for a while. He also fostered a debate on foreign and security policy among intellectuals, which fed into the summer leadership meetings at Beidaihe. One side argued that the Kosovo War and the Belgrade bombing did not represent a significant change in the geopolitical equation, the intentions of the United States, or the context of China's security. They argued that for a relatively weak China to confront the United States would be too dangerous. Another asserted that the United States was "bent on maintaining its global hegemony by military means," and that China should take the lead in organizing a coalition against U.S. hegemonism. In the end, the leadership arrived at a new consensus that emphasized continuity ("peace and development" was still the dominant trend) but which bowed toward the more negative view and admitted that "hegemonism and power politics" was on the rise.

It happened that Lee Teng-hui's July 1999 statement concerning "special state-to-state relations" occurred during the middle of the response to Kosovo and Belgrade. Again, there was an outburst of nationalistic fervor, which continued for a month or so. When the leadership met at Beidaihe for its summer conclave, Taiwan was discussed at length and critics said their piece. Jiang likely reminded them that relations with the United States were on the mend after the damage caused by Belgrade, and that China lacked the ability to respond militarily. Ultimately, however, the meeting "agreed on a nuanced but only slightly less militant approach to the Taiwan issue. Qian [Qichen] announced . . . essentially a reiteration of existing policy." And Jiang secured a symbolic victory at his September meeting with President Clinton at the Asia Pacific Economic Cooperation (APEC) Summit, when Clinton acknowledged that Lee's remarks had created trouble.[15]

Similar dynamics were at play after Lee Teng-hui's visit to the United States, Jiang came under some pressure not only from within the immediate leadership circle but from outside it as well. Leftist forces had been criticizing him publicly over the difficulties facing state-owned enterprises and the growth of private firms. Then Lee's visit sparked a firestorm of nationalistic attacks. This was, for example, the period that the nativist tract, *China Can Say No*, was published. Moreover, elements of the party and the government criticized the Ministry of Foreign Affairs for its weakness over U.S.-China relations. Under this pressure, "those with more moderate views on Sino-U.S. relations found it difficult to express their opinions." Jiang had to bide his time before regaining the initiative.[16] In 2003 and 2004, some intellectuals and chat-room netizens called for a stronger response to Chen Shui-bian's proposal for a new constitution and his re-election.

In sum, Jiang Zemin and his colleagues in the leadership have managed the politics of foreign policy by permitting the controlled expression of conflicting views, some of them fairly critical.[17] His *modus operandi* is consistent with the view of those scholars (Suisheng

Zhao, for example) who believe that the regime uses nationalism pragmatically, in a way that is instrumental, reactive, and state-centered.[18] It is also appropriate for the asymmetrical strategic situation in which the PRC finds itself, with a security culture that favors robust displays of resolve in the face of threats, but a military establishment that does not yet possess the capabilities necessary to successfully engage in coercive diplomacy.

In terms of crisis management, however, a strategy that manipulates nationalism instrumentally by permitting constrained public ventings in times of trouble carries some risks. There was, for example, the danger that the violent demonstrations against the United States in May 1999, which the regime tolerated, might be turned against it.

Cross-Strait (Non)Communication.

The absence of authoritative cross-Strait communication compounds the difficulties of crisis management. Again, this is a complicated subject with a long history. The essential point is that since 1995, with one exception, Beijing has insisted that Taiwan commit to the "one-China principle" before formal dialogue can occur. It does so because it has concerns about Taiwan's fundamental intentions and wants some reassurance that what it believes is goodwill will not be exploited. Yet this stance has a perverse impact. The desire for reassurance on the big picture deprives Beijing of a mechanism to ensure that episodes of tension do not spin out of control.

Taipei is unwilling to accept the one-China principle, for fear of making concessions up front that would sacrifice fundamental interests. And it has reason to be wary. As part of its negotiations playbook, Beijing seeks to get its adversary's acceptance of basic principles at the outset, which it then manipulates to its advantage thereafter. Even without this clear tendency, imposing preconditions for negotiations generally denies the side that is asked to accept them sufficient information about what the end of the process will be. Knowing the end-state is particularly important in the case of Taiwan, for which the issue is the island's ultimate future and where leaders are publicly accountable through elections.[19]

Substantively, Taiwan has had several concerns about the one-China principle. For purposes of cross-Strait relations, it is now defined as follows: "There is only one China in the world. Both the mainland and Taiwan belong to one China. China's sovereignty and territorial integrity brook no division." The Taiwan government would find this formulation problematic on a couple of counts. First of all, it would fear that accepting the idea that China's sovereignty cannot be divided would undermine its fundamental claim that it possesses sovereignty. Second, it would not necessarily regard the statement that the mainland and Taiwan belong to one China as concurrence with the equality it asserts between the Republic of China (ROC) and the PRC. In referring to the mainland and Taiwan, Jiang is citing geographic entities, not governmental ones. And to say that both "belong" implies no equivalence.

Moreover, Taipei is also quite aware that this is the definition of the one-China principle that Beijing uses for cross-Strait relations. There is another one, often used for the international community, that is an even more explicit rejection of Taiwan's view of its legal identity. Its elements are that there is one China in the world, Taiwan is a "province" of that China, and the PRC government is the sole representative of China. This formulation indicates clearly that, in Beijing's eyes, Taiwan is a unit subordinate to the central government and has no equivalent role in the international system. As such, it is a stronger reason to view the one-China principle as a negotiating trap and a reason to refuse to accept it.

There is another formula that Beijing cites as a means to resume dialogue, if only Taipei would reaffirm it. That is the so-called 1992 consensus, the exchange of fax messages by which the two sides agreed to "orally" express their respective views on the one-China principle in order to facilitate the April 1993 meeting between Koo Chen-fu and Wang Daohan and the signing of several technical agreements.[20] In theory, this set of overlapping statements might be a basis for the two sides to return to dialogue, if Beijing needs a symbolic fig leaf to justify resuming dialogue for pragmatic reasons, and if Taipei were willing to take a chance on offering one. But in the absence of a prior consultation to build mutual confidence as to what the "consensus" means, it may have outlasted its value as a means to bridge differences. The two sides now differ considerably on what the consensus means. Beijing asserts, as Qian Qichen put it in January 2003, that "the two organizations reached a consensus allowing each side to express in verbal form the formulation 'both sides of the Strait uphold the one-China principle.'"[21] The most that people in Taiwan would say is that the two sides had their respective ways of defining one China, a view that Beijing has consistently rejected. That being the case, the Chen government feared that accepting the 1992 consensus would end up being a back-door acceptance of the one-China principle as Beijing defines it.

The breakdown of public dialogue has been accompanied by an absence of private contacts. There was at least one instance—in the early 1990s—when the authoritative representatives of the senior Beijing and Taipei leaderships met regularly over a period of 4 1/2 years. Lee Teng-hui's representative was Su Chih-ch'eng, the member of his staff whom he trusted the most. There were a series of interlocutors on the PRC side: Nan Huajin, a Hong Kong figure with good mainland connections; Yang Side, a subordinate of Yang Shangkun, then PRC state president and the key figure on Taiwan policy; Wang Daohan; and Zeng Qinghong. What we know of the operation of this channel suggests key interlocutors were clearly speaking for their principals, so all could have confidence in what was said. Each side could explain the dynamics of its domestic political environment and how they would—or would not—affect negotiations. Each could test its ideas for removing substantive and procedural obstacles. Each could preview upcoming actions and statements, place them in an objective context, and influence the other's reaction. But the contacts ended with Lee's visit to the United States, and as far as is known, have not resumed in any similar form.[22]

Subsequently, there have been any number of individuals—businessmen, scholars, and so on—who act as self-appointed vehicles of private communication. And there are times when one side or the other at least gives the impression that these go-betweens operated with its blessing. All too often, however, these mechanisms fail because intermediaries do not really speak for the leaders they claim to represent. As a result, confidence in these channels has declined over time. Similarly, some individuals in the Democratic Progressive

Party (DPP) have contacts with people on the mainland. Again, there is often uncertainty how effective such channels would be in a crisis.

The upshot is that no authoritative channels currently exist to mitigate tensions and contain conflict when they occur. For Beijing, communication with Taiwan takes a back seat to securing a pledge of the latter's *bona fides*. Rejected is the contrary approach, that communications channels are most required when mutual suspicion is the greatest—the stance that the PRC has taken since October 2002 concerning the United States and North Korea.

The American Factor.

China, of course, regards the United States as a key factor in the Taiwan equation. In January 2001, Qian Qichen offered a telling—and exaggerated—commentary on China's view of the American impact:

> If foreign countries [i.e., the United States] interfere in the Taiwan issue, the local Taiwan Independence factions will rely on this kind of foreign interference to stir up splittism, and cause the Taiwan problem to drag on forever. That is just not possible. The question of national reunification must be decided. If the American Government takes a stance of supporting peaceful reunification, then it will be of very great use. If it says to Taiwan, "do not be afraid, we will protect you and we will sell you arms, we will stand behind you, we can act behind the scenes for you," then the situation is quite different. Consequently, if the U.S. wants to play a role here, first they must not support Taiwan independence, and they must not support Taiwan splitting away. They must not support any type of splittist activities by Taiwan on the international stage. [If they do not,] I do not see where Taiwan has any power, nor any reason to refuse reunification.[23]

That is, Beijing sees American support to Taiwan as both the only obstacle to a successful application of pressure and intimidation and a stimulus for the Taiwan initiatives that, in its eyes, reflect a looming separatist break-out. By implication, energizing the United States to block those initiatives has been an essential element in Chinese crisis management. And in each of our four episodes, Beijing both blamed Washington to some degree for encouraging Taipei's behavior and sought to get the United States to put it back in the box.

In 1995-96, China concluded that the United States betrayed an absolute obligation to deny Lee a visa and was therefore complicit in his effort. Once the visit occurred, the PRC then suspended various unrelated elements of the bilateral relationship and used displays of force to challenge American credibility. There then ensued a lengthy process, over 2 years *in toto*, of putting the relationship back together. The culmination of that process, as far as Taiwan was concerned, was Clinton's "three nos."

In 1999, Washington was concerned enough about the PRC reaction (and specifically that an incident might result from ROC Air Force (ROCAF) and PLAAF planes flying close to each other) that it dispatched me to Taiwan and Stanley Roth and Ken Lieberthal to Beijing to urge restraint on both sides. That concern continued until September and the APEC meeting, when Clinton told Jiang Zemin that Lee's words had "caused trouble." China portrayed this as an American judgment that Lee was a troublemaker.

In early 2000, the Clinton administration was very worried that Beijing's intemperate remarks—both in the February white paper and Zhu Rongji's March warning—reflected a move towards some sort of military action. Again to encourage restraint, it sent Lee Hamilton and me to Taipei and used the previously scheduled Beijing trips of Richard Holbrooke and Sandy Berger. Washington remained firm in its opposition to the use of force, and added a new rhetorical requirement, that the Taiwan Strait issue be resolved with the assent of the people of Taiwan.

In late 2003, after keeping a low profile in the Taiwan election campaign, Beijing temporarily abandoned its policy of restraint. On the one hand, it sought to reassert the credibility of its resolve. PRC military experts warned that losing Taiwan was not an option. China was prepared to pay a high cost to "oppose Taiwan independence," including giving up the Olympics, economic modernization, and so on.[24] On the other, it intensified pressure on the Bush administration to restrain Chen. (In fact, Washington was itself frustrated that Taipei had ignored its repeated requests for restraint and had not engaged in the kind of prior consultation on initiatives that might provoke a forceful PRC response and draw the United States into a conflict by virtue of its security commitment to Taiwan and the need to preserve its own credibility.) Frustration and concern culminated in President Bush's public statement on December 9, 2003, in the presence of PRC Premier Wen Jiabao that, "We oppose any unilateral decision by either China or Taiwan to change the status quo. And the comments and actions made by the leader of Taiwan indicate that he may be willing to make decisions unilaterally to change the status quo, which we oppose."

Washington's default approach on the Taiwan Strait issue is one of dual deterrence, combining warnings and reassurance to both sides of the Strait. On the one hand, it warns Beijing not to use force at the same time that it reiterates American nonsupport (or even opposition) to Taiwan independence. On the other, it warns Taiwan not to take political initiatives that might provoke a violent response from China, while reaffirming its support for Taiwan's security. Beijing's strategy in times of stress is to maximize Washington's reassurance to it and encourage it to constrain Taipei more effectively. And it showed a willingness, particularly in the early episodes to use displays of force to demonstrate its resolve and, perhaps, rattle the United States into restraining Taipei. Moreover, the fact that the PRC usually does not have authoritative communications channels with Taiwan increases the American role.

Summing Up.

There remains much that we do not know in the United States about how China manages the Taiwan Strait issue under conditions of stress. We would understand much more if we had the fine detail of leadership behavior, such as that which has been available for decades on the Kennedy administration's actions during the Cuban missile crisis. Also valuable would be more information on the operational dimension of crisis management, the difficulty of which has become apparent in the post mortems on the Clinton and Bush administrations' struggle to meet terrorist threats.

What seems clear is that China approaches the Taiwan Strait issue fearful that the island's leaders are going to "break out" and permanently separate Taiwan from China, thus posing a challenge to its fundamental interests in national unification. Aggravating this fear, I argue, is a Chinese misperception of Taipei's intentions. In situations that qualify as crises, China's decisionmaking system is prone to become even more centralized and personalized than it normally is, and to overreact to perceived Taiwan provocations. Political dynamics are mixed. Within the leadership, the Taiwan issue has not created factional splits, but it does not lend itself to moderate responses when tensions are high. With respect to the broader public, the leadership has shown some skill in managing nationalistic sentiment, but at some risk. The absence of direct dialogue makes crisis management more difficult. And Beijing relies on the United States to restrain Taiwan politically so it does not have to act militarily.

Has China learned anything from the episodes that occurred over the last 9 years? There is, it appears, a contrast between Beijing's response to Lee Teng-hui's American visit and his "state-to-state" proclamation on the one hand, and the stance taken toward the 2000 and 2004 elections. The former were met with displays of force, the latter with talk of force ("Taiwan independence means war"). And from 1996 to 2000 to 2004, China took a progressively more restrained approach to the island's presidential elections. China gained, it is clear, a growing understanding that taking an aggressive stance during the election campaign would inflame Taiwan opinion and bring about the very result that it sought to prevent.[25] We can debate, however, whether this greater moderation in the attempted exercise of Chinese influence reflects learning, or was simply a function of the political strategy that Beijing adopted after May 2000. That strategy was to rely on economic attraction between the two sides of the Strait and the pan-Blue opposition within Taiwan to make Chen Shui-bian a one-term president. Whatever the case, the lesson for China of the March 2004 election, in which the campaign skill of the pan-Green overcame the pan-Blue's critiques of Chen's performance, would be that moderation does not work.

If we look at the six factors discussed above, and granting that managing situations of stress is not easy and that other actors do have their own deficiencies, the future as it applies to Beijing is not comforting:

- China's sense of threat is probably more profound, as its confidence in the pan-Blue as a check against Chen has declined. Although some in China are studying creative alternatives to long-standing policy, the more dominant voice emphasizes the need for coercive capabilities.

- Similarly, Beijing's assessment of Chen Shui-bian's and Lee Teng-hui's intentions has not changed ("It's independence, stupid!"). It has no trust in his recent inaugural commitments concerning constitutional revision.

- Although the membership of the Taiwan Affairs Leaders' Small Group has turned over, it is not clear whether the Fourth Generation's views are any more creative than those of Jiang Zemin's cohort, or whether it will reduce the tendency of the system to over-react when it comes to perceived Taiwan challenges.

- Politically, Taiwan is still the third rail and nationalistic sentiment appears to remain strong.

- Whereas a way was found after 1996 to resume dialogue, Beijing missed a significant opportunity in 2000 to engage Chen Shui-bian. How it did so made Chen more cautious. Having set Chen's acceptance of the one-China principle as the precondition for dialogue in 2000 and so backed itself into a corner, China will find it very difficult to set that precondition aside in 2004 and beyond.

- Finally, the United States remains deeply involved in ensuring that those crises or mini-crises that occur do not spin out of control. In Beijing's mind, Washington's role has become even more significant as hope for the pan-Blue recedes.[26]

In short, there is little reason to believe that China's handling of future situations of stress concerning Taiwan will be any better than its performance in the past. To be sure, China has many reasons not to go to war over Taiwan, and the leadership understands them. Yet the leadership also believes that there are certain lines that Taipei cannot be allowed to cross. The process by which will weigh these competing interests in the future will itself affect the outcome.

ENDNOTES - CHAPTER 5

1. This chapter draws extensively from my volume on cross-Strait relations, to be published in the summer of 2005 by the Brookings Institution Press and entitled *Untying the Knot: Making Peace in the Taiwan Strait.*

2. Thomas J. Christensen, "The Contemporary Security Dilemma: Deterring a Taiwan Conflict," *Washington Quarterly*, Vol. 25, No. 4, Autumn 2002, pp. 12-13.

3. Some would argue that Chen went further than Lee in a separatist direction. They certainly spoke of these issues in somewhat different ways. Yet the core of their thinking—that the entity over which they presided possessed sovereignty and that any unification had to begin on that premise—was the same. Their differences were largely a function of where they started out (Lee in an anti-independence party and Chen in an anti-unification one), and the political constituencies to which they therefore had to appeal.

4. *Zhongguo Shibao* investigative report, *Zhongguo Shibao*, December 12, 2000 (translation in author's files); "Li Teng-hui Defends State-to-State Relationship," *Zhongguo Shibao*, September 1, 1999, *Foreign Broadcast Information Service* (FBIS), CP9909010005, accessed September 1, 1999; interview with Lin Bih-jaw, February 24, 2003. It is interesting to speculate whether the announcement that PRC leaders reportedly intended to make was, in fact, an earlier version of "The One-China Principle and the Taiwan Issue," the "White Paper" ultimately issued in February 2000.

5. Michael D. Swaine, "Chinese Decisionmaking Regarding Taiwan, 1979-2000," in David Michael Lampton, ed., *The Making of Chinese Foreign and Security Policy in the Era of Reform*, Stanford: Stanford University Press, 2001, pp. 307-309.

6. Swaine, "Chinese Decisionmaking Regarding Taiwan," p. 309; Ning Lu, *The Dynamics of Foreign-Policy Decisionmaking in China*, Boulder: Westview Press, 1997, p. 76.

7. Swaine, "Chinese Decisionmaking Regarding Taiwan," p. 309.

8. See, for example, Robert L. Suettinger, *Beyond Tiananmen: The Politics of U.S.-China Relations, 1989-2000*, Washington, DC: Brookings Institution Press, 2003, pp. 200-263; John W. Garver, *Face Off: China, the United States, and Taiwan's Democratization*, Seattle: University of Washington Press, 1997; the authors in Suisheng Zhao, ed., *Across the Taiwan Strait: Mainland China, Taiwan, and the 1995-1996 Crisis*, New York: Routledge, 1999, particularly You Ji, "Changing Leadership Consensus: The Domestic Context of War Games," pp. 77-98; Swaine, "Chinese Decisionmaking Regarding Taiwan"; Ross, "The 1995-96 Taiwan Strait Confrontation," *International Security*, Vol. 25, No. 2, Fall 2000, pp. 87-123; Allen S. Whiting, "China's Use of Force, 1950-1996, and Taiwan," *International Security*, Vol. 26, No. 2, Fall 2001, pp. 120-123; and Tai Ming Cheung, "Chinese Military Preparations Against Taiwan Over the Next 10 Years," in James R. Lilley and Chuck Downs, *Crisis in The Taiwan Strait*, Washington, DC: National Defense University Press, 1997.

9. Ross, "The 1995-96 Taiwan Strait Confrontation," p. 118. For a more recent assessment of the 1995-96 events by Major General Huang Bin of China's National Defense University that emphasizes the benefits, see "Peace Will Prevail Only When the Two Sides of the Taiwan Strait Are Reunified—an Interview with Maj. Gen. Huang Bin, Noted Military Strategist on the Mainland," *Dagongbao*, May 13, 2002, FBIS, CPP20020513000048, accessed February 28, 2004.

10. For an example of this restraint in operation, one in which PRC Taiwan specialists argued for "calmness" in Beijing's response to Chen Shui-bian's August 2002 statement that there were "two countries on each side" of the Taiwan Strait, see "Roundup: Chen Shuibian's Blunt Remarks Stir Up Turmoil, Mainland Experts Call on Making a Counterattack with Rationality and Calmess," *Zhongguo Tongxunshe*, August 9, 2002, FBIS, CPP20020809000090. Presumably, the scholars' public advice was a similar to what they were telling officials. And by inference, the officials listened.

11. Robert L. Suettinger, "China's Foreign Policymaking Process," paper prepared for a Center for Strategic and International Studies meeting on "PRC Policymaking in the Wake of Leadership Change," October 31, 2003.

12. Suettinger, *Beyond Tiananmen: The Politics of U.S.-China Relations, 1989-2000*, pp. 224-225, 245-246, 262-263; You Ji, "Changing Leadership Consensus"; Garver, *Face Off*; Cheung, "Chinese Military Preparations"; Michael Swaine, "Decisionmaking Regarding Taiwan," pp. 319-327; Jing Huang "The PLA's Role in Policymaking Under the New Leadership," paper prepared for a session on "PRC Policymaking in the Wake of Leadership Change," Center for Strategic and International Studies, October 31, 2003, p. 13. David Finkelstein, in his paper for the same conference, states that the precise PLA role is basically unknown; see his "The Role of the PLA in the Chinese Policymaking Process."

13. "Jiang Puts Hard Line to the Test in China, *Washington Post*, May 31, 2004.

14. David M. Finkelstein, "China Reconsiders Its National Security: The Great Peace and Development Debate of 1999," Project Asia, The CNA Corporation, December 2000.

15. Suettinger, *Beyond Tiananmen*, pp. 384-385.

16. Joseph Fewsmith and Stanley Rosen, "The Domestic Context of Chinese Foreign Policy: Does Public Opinion Matter?" in David Michael Lampton, ed., *The Making of Chinese Foreign and Security Policy in the Era of Reform*, Stanford: Stanford University Press, 2001, pp. 176-177.

17. Finkelstein, "China Reconsiders Its National Security," p. 6.

18. Zhao, "Chinese nationalism and its International Orientations," p. 23.

19. Richard H. Solomon, *China's Negotatiating Behavior: Pursuing Interests Through 'Old Friends'*, Washington, DC: United States Institute for Peace Press, 1999, pp. 71-75.

20. The Association for Relations Across the Taiwan Straits (ARATS) said, "Both sides of the Taiwan Strait uphold that One-China principle and strive to seek national unification. However, in routine cross-Strait consultations, the political meaning of 'One China' will not be touched upon." The Straits Exchange Foundation (SEF) said, "Although the two sides uphold the One-China principle in the process of striving for cross-strait national unification, each side has its own understanding of the meaning of one-China."

21. "Apparent Text of Qian Qichen Speech at Forum on Jiang's Eight-Point Proposal," *Xinhua Domestic*, January 24, 2003, FBIS, CPP20030124000041, accessed January 24, 2003.

22. For this channel, see Zou Jingwen, *Li Denghui Zhizheng Gaobai Shilu* (Record of Revelations on Lee Denghui's Administration), Taipei: INK, 2001, pp. 194-197.

23. "Shemma Wenti Dou Keyi Taolun: Qian Qichen Fuzongli Jieshou Huashengdun Youbao Jizhe Pan Wen Caifang" (Anything Can Be Discussed: Vice Premier Qian Qichen Accepts an Interview by John Pomfret of the *Washington Post*), *Shijie Zhishi*, 2001, No. 3, February 2001, p. 14.

24. Yang Liu, "Military Experts on War to Counter 'Taiwan Independence': Six Prices; War Criminals Cannot Escape Punishment," *Renmin Wang*, December 3, 2003, FBIS, CPP20031203000036, accessed March 27, 2004.

25. And at least one PRC scholar, Xu Shiquan, believes that Zhu Rongji's bristling prediction did not help Chen win in March 2000.

26. It was after the pan-Blue caved on the referendum issue in November 2003 that China escalated its pressure on the United States.

CHAPTER 6

DECISIONMAKING UNDER STRESS:
THE UNINTENTIONAL BOMBING OF CHINA'S BELGRADE
EMBASSY AND THE EP-3 COLLISION

Paul H. B. Godwin

INTRODUCTION

The accidental bombing of China's Belgrade embassy in 1999 and the 2001 collision between a U.S. Navy (USN) reconnaissance aircraft and a Chinese navy fighter over the South China Sea were serious incidents in Sino-American relations. Both resulted in Chinese and American leaderships making decisions under high stress. At a simple descriptive level of analysis, both events were potentially avoidable tragedies. In the night of May 7, 1999 (early morning of May 8 in China), a faulty target designation process resulted in a U.S. Air Force (USAF) B-2 bomber unintentionally striking China's Belgrade embassy with three global positioning system (GPS)-guided 2,000 lb bombs. Three Chinese journalists were killed and 20 embassy staff injured, together with extensive damage to the embassy compound. The collision between a USN EP-3 electronic surveillance aircraft and a People's Liberation Army Navy (PLAN) F-8 on April 1, 2001, resulted from one of the two F-8 interceptors making an error as it maneuvered around the EP-3. The Chinese aircraft disintegrated and crashed into the South China Sea. The pilot was never recovered. The EP-3 was so extensively damaged, it had to make an emergency landing at the People's Liberation Army Naval Aviation (PLAN-Av) Lingshui airfield on Hainan Island, where the aircraft's 24 crew members were detained for 11 days. In both cases, human error rather than a deliberate policy choice was the incident's catalyst.

This chapter will assess these incidents as case studies of Chinese and American decisionmaking under stress. The assessment will be structured into five components. First, the "context" will evaluate the state of Sino-American relations at the time of the incident. Second, the "response" will be assessed by evaluating the manner in which

161

both governments reacted to the incidents. Third, the "negotiations" will be evaluated to determine what each government sought to achieve and tried to avoid as they negotiated a resolution to the incidents. Fourth, the "resolution" will be assessed to determine to what extent each government achieved its negotiating objectives. Finally, the negotiating strategy employed by Beijing will be assessed to determine whether there are patterns of decisionmaking under stress that are potentially useful for predicting China's behavior in any future incidents producing high levels of tension in Sino-American relations.

THE UNINTENTIONAL BOMBING

Context.[1]

Sino-American relations were rapidly deteriorating when the USAF B-2 unintentionally bombed China's Belgrade embassy. The immediate cause of the degeneration was China's objection to the North Atlantic Treaty Organization's (NATO) air attack on Yugoslavia. NATO's decision to launch Operation ALLIED FORCE was not taken to the United Nations Security Council (UNSC) for consideration and was therefore in Beijing's eyes an illegitimate use of military force. China's concerns, however, ran deeper than what Beijing saw as unilateral action by a U.S.-dominated alliance.

Following the Taiwan Strait confrontation of 1995-96, President Jiang's 1997 visit to the United States and President Clinton's reciprocal visit to China in 1998 had significantly improved Sino-American relations. Nonetheless, Beijing's longstanding concerns about U.S. post-Cold War strategy and intentions toward China remained.[2] The core issue that arose was whether Deng Xiaoping's "peace and development" principle declared in 1985 remained valid. Deng had argued then that the primary concern of the world's major powers was peace and economic development. By the late 1990s, some Chinese analysts were suggesting that U.S. post-Cold War foreign policy had negated Deng's prediction and China's security was now threatened by American "hegemonism and power politics." In addition to the Kosovo intervention, those arguing that American interventionism and power politics were a trend that threatened China's security had much to draw upon. The United States was strengthening its alliances in Europe and Asia, and in recent years had intervened in Panama, Haiti, Somalia, and Bosnia. Furthermore, the United States was legitimizing its "interventionism" by declaring that gross violations of human rights required the international community to override the sovereign right of a state to govern within its own borders. NATO had applied this rationale for its bombing campaign on Yugoslavia.[3] In essence, Beijing argued that as the world's sole superpower, U.S. "hegemonism" and "power politics" were the prime source of the threat to global stability and to China.

Furthermore, U.S. domestic politics made Beijing fully aware of the anti-China disposition in the Congress and its support for Taiwan.[4] The Defense Authorization Act of 1999 authorized the Secretary of Defense to study the theater missile defense (TMD) architecture that would be required to defend U.S. allies in the Asia-Pacific region. Taiwan was not mentioned, but it was evident Taipei was included as an ally to be defended. In part responding to domestic political pressures, in 1999 China's human rights deficiencies returned as a major irritant in Sino-American relations. The U.S. sponsored a resolution criticizing Beijing's human rights record at the UN Human Rights Commission—something it had not done in 1998. These developments were compounded by a *New York Times* story in March originating in the Cox Committee.[5] The *Times* reported the committee as concluding that China had stolen a nuclear weapon design from one of the Department of Energy's laboratories, most likely Los Alamos. This story revived past allegations of Chinese illegal contributions to political campaigns. These events clearly endangered the President's objective of working toward a "strategic cooperative partnership" with China announced during President Jiang Zemin's 1997 visit to the United States. What is more, even as President Clinton's China policy was under severe attack his presidency was weakened by personal failings that led to his impeachment, making his commitment to a partnership with China even more endangered.

Despite internal disagreement, but seeking to improve Sino-American relations, the Clinton administration revived consultations with Beijing over China's accession to the World Trade Organization (WTO). Some in Beijing were cautious or opposed because they feared the consequences of opening up China's economy. Premier Zhu Rongji, however, was confident that China's economy could withstand the opening up WTO would require and that agreement could be reached.[6] Negotiations between the United States and China were undertaken in Beijing meetings held in March, and Premier Zhu planned a visit to the United States for April. Unfortunately, the timing of his visit coincided with the opening phase of NATO's air assault on Serbia, which commenced in the night of March 24-25, during President Jiang Zemin's state visit to Italy. Some of the attacking aircraft were launched from Italian bases.

With NATO bombing Serbia, there was growing opposition in Beijing to Zhu Rongji's visit to the United States. Zhu himself was reluctant. Upon returning to China, Jiang Zemin called a Politburo meeting to resolve this issue. It was decided that the benefits of maintaining a working relationship with the United States, together with WTO membership, outweighed the potential costs, so Premier Zhu's visit should proceed as scheduled. Nonetheless, it was also agreed that China would increase its criticism of the United States and NATO.[7] As it transpired, disagreement within the Clinton administration and the President's concern that political opposition to China in the Congress was too intense caused President Clinton to conclude on April 7 that he could not support China's accession. Despite the concessions he brought with him to Washington, Zhu Rongji returned to Beijing empty-handed and subjected to severe criticism. His failure became the kindling for the dissatisfaction within China's leadership over the Sino-American relationship.

Response.

Beijing's reaction to the bombing of its Belgrade embassy in the morning of May 8 (May 7 in Europe) was one of stunned incredulity. China could not believe the bombing was an accident. It touched off an angry, sharp, and deeply felt patriotic reaction across China, especially among students. Moreover, coming as it did on the heels of Zhu Rongji's failure to get the expected agreement on WTO accession and in the midst of a major deterioration in Sino-American relations, it unleashed even more criticism of Premier Zhu Rongji. Internet chat rooms charged him with being a "traitor." Jiang Zemin's past concern with sustaining a working relationship with Washington exposed him to the criticism that he was "soft" on the United States.[8]

At a Politburo meeting in the morning of May 8, Jiang Zemin faced a dilemma: How not to sacrifice what he saw as the need to sustain a pragmatic relationship with the United States while simultaneously responding effectively to the outpouring of nationalism and patriotism unleashed by the bombing.[9] For Jiang and other Chinese leaders, the importance of Sino-American relations was linked to China's strategic objectives. Trade, technology transfers, and investment from the United States were critical for China's economic development. Jiang therefore had to respond to the bombing in such a way as to be seen as tough on the United States by his critics, but without totally undermining the relationship itself. It was also useful to Jiang Zemin that the patriotic anger of students could be directed at the United States. With the 10th anniversary of the PLA's crackdown on student demonstrators in Tiananmen less than 1 month away, it was far better that their anger now be focused on the United States.

After considerable debate, the Politburo made the following decisions: 1. To condemn the bombing and summon the U.S. Ambassador and charge him with delivering China's strongest protest to NATO; 2. Demand a special meeting of the UNSC to discuss and denounce the bombing; 3. Dispatch an aircraft to Belgrade to bring back the appropriate embassy personnel; 4. Provide guidance for the conduct of organized protests at U.S. diplomatic facilities across China; and, 5. Ensure that there were sufficient public security personnel to prevent any extremist behavior and to maintain public order during the demonstrations.[10]

In the morning of May 8, Ambassador Sasser reached China's Ministry of Foreign Affairs (MFA) to apologize for the tragic error and offer his condolences. Later in the day, angry crowds began gathering around the American Embassy as news accounts of the bombing and TV coverage of Yugoslav rescue teams comforting the wounded and weeping survivors enraged Chinese viewers even more. The MFA described the bombing as barbaric and summoned

Ambassador Sasser to receive China's protest. That afternoon, President Clinton met with reporters where he declared the bombing to be a tragic mistake and expressed his condolences to the Chinese people.

Although transported to the embassy area and watched by police and soldiers, the anger of the university students, like all the others gathering to protest the bombing, was clearly genuine.[11] What does seem to be in accord with observers on the scene is that China's authorities were regulating and using this anger. That evening, however, as busses returned students to their campuses the crowd grew larger and harder to control. Rocks and paint bombs were thrown over the embassy walls. More dangerous were the occasional Molotov cocktails. To the ambassador and embassy staff, it appeared that, despite the presence of large numbers of police, the mob was out of control. Early Sunday morning, May 9, fearful that the unruly horde outside could break into the compound, embassy staff began destroying sensitive documents.[12]

On the same day, President Clinton sent a letter to President Jiang Zemin expressing his apologies and condolences. The day before, Secretary of State Madeline Albright had personally delivered a letter to the Chinese embassy for Foreign Minister Tang Jiaxuan. Beyond their apologies and condolences, however, both letters declared the bombing campaign was warranted and would be sustained.[13]

Although it is unclear whether the action was in response to these letters, on Monday, May 10, *Xinhua* reported that China was suspending military contacts with the United States and postponing consultations on human rights, arms control, international security, and nonproliferation.[14] May 10 also saw China's Foreign Ministry make its demands to the United States. These were a formal public apology to the Chinese government and people and to the families of the victims of the bombing; a complete investigation of the bombing with prompt disclosure of the findings followed by severe punishment of those responsible. Beijing sought international pressure on the United States by requesting a UNSC meeting to condemn the bombing while threatening to veto any UN sponsored peace plan for Yugoslavia unless the bombing was terminated.[15]

As these actions were underway, the government's channeling of the public's fury continued. A speech by Vice-President Hu Jintao broadcast on Sunday afternoon reflected the dual concerns of Jiang Zemin that he must be seen as tough on the United States, while not completely disrupting relations with Washington. Hu condemned U.S. and NATO's bombing and praised the patriotism of the crowds protesting around the embassy compound and American diplomatic facilities across China. He equally emphasized the need to be orderly and not to overreact and disturb social order. On Monday, May 10, the message was somewhat different. The morning newspapers charged the United States with criminal acts and seemed designed to increase public anger. Students were again bussed to the embassy area, as were government employees. Now, however, they were required to show proof they had permission to demonstrate and were closely supervised. Thousands marched on Monday, but the numbers dwindled to a few hundred on Tuesday, and on Wednesday the demonstrations ended.[16]

On Thursday, May 13, Jiang Zemin headed a leadership meeting honoring the returning bodies of those killed in the May 7 bombing, whom he declared "martyrs." Jiang's speech[17] reflected his effort to compromise on those aspects of Sino-American relations dividing China's leadership. He sought to be tough on the United States and a supporter of the patriotism reflected in the demonstrations around the embassy compound and across the country, while also stressing the continuity in China's domestic and foreign policies. The demonstrations, he said, had confirmed the cohesion of the nation and that China cannot be bullied. The embassy bombing was declared to be a "brutal act" and the United States was charged with using its "economic, scientific, technological, and military prowess" to "practice hegemonism and power politics" and interfere in the internal affairs of other countries. Nevertheless, Jiang stressed that building China's strength was the country's "central task." This required continuing the policy of "opening wider to the outside world" and maintaining "social and political stability and unity." Over succeeding days, *People's Daily* articles suggested that Jiang's speech did not reflect a leadership compromise. Rather, that the division over China's policy toward the United States continued. Articles reflecting quests for both a hard line, confrontational approach to United States and for sustaining a pragmatic policy were evident.[18]

At the request of China's UN Mission, the UNSC met to consider the bombing. On May 14, the UNSC observed a minute of silence and expressed its "profound regrets" over the bombing and its "deep sorrow" over the lives lost, injuries, and property damage to China's Belgrade embassy. It also took note that "regrets and apologies were expressed" by NATO's members, and that NATO had initiated an investigation of the bombing.[19] The same day saw President Clinton call President Jiang Zemin to personally apologize and repeat that a thorough investigation was underway and that its findings would be fully disclosed.

Negotiations.

Continued division within the Chinese leadership probably explains why Under Secretary of State Thomas Pickering was unable to promptly visit China and provide the official and very detailed explanation of what had led to the unintentional bombing. Secretary Pickering was prepared to travel with his delegation in late May or early June, but Beijing delayed the trip on the grounds that public opinion was yet too intense for such a visit. On June 12, however, Prime Minister Qian Qichen declared that "China does not want a confrontation with the United States."[20] On June 17, accompanied by a delegation of high-level officials from the White House, the Department of State, Department of Defense (DoD), and the Intelligence Community, Secretary Pickering was finally allowed to present his briefing to Foreign Minister Tang Jiaxuan.[21] President Jiang Zemin, however, had departed Beijing, presumably to avoid meeting the U.S. Undersecretary of State.[22] Not unexpectedly, China rejected the explanation.[23] Moreover, despite the Central Intelligence Agency's firing in April 2000 of one unnamed employee and taking disciplinary actions against six others for the targeting errors, the Chinese government has yet to accept Washington's explication.[24]

Resolution.

Nonetheless, when preparing to leave his post at the end of June 1999, Ambassador Sasser met separately with President Jiang Zemin, Premier Zhu Rongji, and Minister of Defense Chi Haotian.

In the Ambassador's view, Beijing was signaling that China wanted to get Sino-American relations back to a more normal footing.[25] Further easing the tensions, in late July after 3 days of negotiations, the United States agreed to pay U.S.$4.5 million in damages to the victims of the bombing.[26]

The months following Secretary Pickering's unsuccessful mission to Beijing were a mix of continued degradation in Sino-American relations and easing of the tensions generated by NATO's attack on Yugoslavia and the embassy bombing. In July, President Lee Teng-hui strained the continuing tensions between Taipei and Beijing even further. In an interview with reporters from a German radio station, he declared that cross-Strait relations were now "at least a special state-to-state relationship," and that Taiwan no longer claimed sovereignty over all Chinese territory. "Consequently, there is no need to declare Taiwan independent."[27] Beijing's suspicions immediately turned to Washington, which then took great pains to reiterate that the United States maintained its "one China" policy, including a telephone call from President Clinton to Jiang Zemin. As Beijing's rhetoric and military threats escalated, the precedent of China's use of military coercion in the 1995-96 confrontation raised concern over a new flare-up. American delegations were quickly sent to Taiwan and China in largely successful efforts to temporarily ease the growing tensions.

With the relaxation of cross-Strait tensions, Sino-American relations began wending their way back toward a more normal pattern. In late September, President Clinton met with President Jiang during the Auckland, New Zealand, sessions of the Asia-Pacific Economic Cooperation (APEC) forum. Clinton once more assured Jiang Zemin that the U.S. "one China" policy was firm, although he also cautioned China against the use of force, and used the meeting to reopen negotiations with Beijing for China's WTO accession. These negotiations were to be no easier than the earlier efforts, but on November 15, 1999, agreement was reached.[28] On November 20, Deputy Assistant Secretary of Defense Kurt Campbell held discussions in Beijing with his Chinese counterparts to reopen military-to-military contacts and restart the Defense Consultative Talks (DCT) suspended by China in response to the bombing.

Improving relations eased negotiations over compensation for the physical damage to China's Belgrade embassy and U.S. diplomatic and consular facilities in China. On December 16, after five rounds of discussions, Beijing and Washington arrived at a compensation agreement. The United States would seek funding from Congress to pay damages of $28 million to China in the State Department's fiscal year 2001 budget. China agreed to pay $2.87 million for the damage to American diplomatic facilities.[29]

Decisionmaking Under Stress.

The unintentional bombing of China's embassy arguably brought Sino-American relations to their lowest point since the June 1989 brutal repression of the Tiananmen demonstrations. What is more, the bombing occurred at a time when relations were already deteriorating with severe divisions within both China and the United States over their future policies toward each other. For Jiang Zemin, seen within his own government as "soft" on the United States, the bombing demanded a circumspect response. He had to effectively respond to the eruption of emotional patriotic anger by being perceived as tough without completely forfeiting the advantages of a pragmatic relationship with the United States. For Jiang Zemin, the most difficult problems created by the bombing were the internal political ramifications. The bombing had served as a catalyst to reignite leadership disagreements over major domestic and foreign policy issues, including the appropriate policy to pursue toward the United States.

For the United States, the bombing was a profoundly humiliating tragedy. Who could believe that, with its vaunted technological superiority, the United States had misidentified a target for precision bombing in a city its diplomatic and military personnel knew well? Perhaps forgotten was the accidental shooting down of an Iranian commercial airliner in 1988 by the USN cruiser *Vincennes*. Human error had in that case interacted with the world's most technologically advanced air defense system to create a tragic mistake. As with Iran in 1988, the U.S. objective was to have its apology and acknowledged responsibility for the tragedy accepted by the Chinese government together with a compensation agreement. Quick achievement of these objectives would best serve the United States.

Whereas the United States sought the quickest resolution possible, disagreements within China's senior leadership and the emotional public response to the bombing meant that President Jiang Zemin would be best served by stretching out the resolution process. This would demonstrate his willingness to be tough in dealing with the United States. U.S. interest in quickly resolving the incident therefore served Jiang Zemin well. Jiang could end the process at a time when he had satisfied the internal political requirement that the United States be handled with resolute toughness but before Sino-American relations were significantly harmed. This strategy appears to be reflected in the events following the bombing and the emotional outburst of patriotism. Rather than seeking a quick resolution, Beijing dragged out the process while simultaneously signaling that China did not want a confrontation with the United States. This strategy can be seen in the high-level meetings granted Ambassador Sasser in June as his posting was coming to an end; in the September Clinton-Jiang summit on the fringes of the Auckland APEC meetings; and in the agreement to reopen WTO negotiations. In short, Jiang Zemin's strategy was to make the United States the petitioner in an extended resolution process. In this, he was successful.

THE EP-3 COLLISION

Context.

Although tension from the unintentional bombing of China's embassy was receding, the final year of the Clinton administration was an uneasy period for Sino-American relations. On the one hand, Beijing and Washington were attempting to restore relations to something approaching normalcy. On the other, Beijing produced two policy papers Washington found disturbing. A white paper on cross-Strait relations issued in February 2000 appeared to lower the threshold for China to use force against Taiwan.[30] Beijing's *Defense White Paper* issued in October identified the United States as the principal threat to global stability and China's security.[31] Governor

Bush's presidential campaign in 1999-2000 added only more uncertainty to the future of bilateral relations by identifying China as a strategic competitor and criticizing the Clinton administration's policy of working toward a strategic partnership with Beijing.[32]

Under these uneasy conditions, progress toward normalcy was made. The Defense Consultation Talks were restored with General Xiong Guangkai's visit to the United States in January 2000. It was agreed that military-to-military contact would be reinstated including high-level meetings, professional military education exchanges, and PLA participation in U.S. sponsored multinational military forums. This agreement spurred a series of high-level military and DoD officials traveling to China for discussions on security and other issues of mutual concern. Port calls to Hong Kong and Shanghai were restored. President Clinton met with President Jiang Zemin in September while both were in New York for the UN Millennium Summit. During his visit, President Jiang gave a luncheon address to business executives and foreign affairs experts. Later in September, President Clinton signed the bill granting China permanent normal trade relations (PNTR) with the United States—legislation that had required considerable political effort to pass through Congress. In November, Clinton and Jiang met again in Brunei where both were attending the APEC leaders meetings. They agreed to reopen the suspended human rights dialogue.

Even as these and other steps toward normal bilateral relations were undertaken, the underlying tensions remained. Most were those that had dogged Sino-American relations for many years. On the U.S. side, human rights, proliferation, and Beijing's refusal to forgo the use of military force against Taiwan and the lack of transparency in defense policies and modernization were at the top of the list. For China, it was arms sales to Taiwan, national and theater ballistic missile defense programs, and what Beijing saw as U.S. "hegemonism and power politics" joined with a "cold war mentality." Beijing's October 2000 *Defense White Paper* added U.S. "interventionism" to these concerns. This pattern of U.S. policies led Beijing to conclude that it was necessary "to enhance its capability to defend its sovereignty and security by military means."[33]

Even with these continuing underlying frictions, when the Clinton administration ended Sino-American relations could be defined as approaching "normal." For some years, normal bilateral relations had encompassed the mutual suspicion and distrust with which Beijing and Washington viewed each other. Nevertheless, the United States and China each had pragmatic reasons for avoiding confrontation and cooperating wherever possible.

This pragmatism can be seen in Secretary of State Colin Powell's January 2001 confirmation hearings and in President Jiang Zemin's *Washington Post* interview 2 months later. In his testimony before the Senate Foreign Relations Committee, Powell stated that although China was not a strategic partner, neither was it an enemy. He defined China as "a competitor, a potential regional rival, but also a trading partner willing to cooperate in areas where our strategic interests overlap."[34] Jiang Zemin's interview was conducted in Beijing shortly after Vice Premier Qian Qichen visited the United States to become acquainted with the new administration. In answering a reporter's question, Jiang Zemin observed, "I don't have a naïve or romantic view that the strategic partnership proposed by President Clinton was a relationship free from struggle or containment. It involved both. Conversely, I do not believe the competitor President Bush talked about does not contain any element of cooperation."[35] Thus, the unfortunate collision between a USN EP-3 and PLAN aircraft occurred at a time when the new Bush administration was clarifying its view of China and the Chinese government was making its adjustments to the new president and his senior advisers.

There were, however, a series of events central to the collision that did not affect high-level Sino-American relations until the incident occurred. For almost a year before the collision, tension over U.S. reconnaissance flights had been intensifying.[36] There had been a candid exchange over this issue during the May-June 2000 Honolulu meetings of the Military Maritime Consultative Agreement (MMCA). Beijing used the MMCA for its complaints because it had been established to prevent military accidents and misunderstandings.[37] A Chinese military official stated that the frequency of American reconnaissance patrols had increased in the latter part of 2000 to four or five times a week some 50 miles off China's coast. In 1997-99, he said, the average number of patrols had been 200 flights a year. The Chinese delegation complained that the patrols were coming too close to China's coastline, and this could cause trouble. The senior U.S. officer at the meeting, Lieutenant General Michael Hagee, USMC, confirmed China's protest. General Hagee recalled replying, "It is international airspace, and we have no intention of modifying what we are doing."

China responded to the increased reconnaissance patrols by conducting more aggressive interceptions. A U.S. Navy official stated that the intercepting fighters had started flying closer and closer to the patrolling aircraft. On Christmas Day, a Chinese fighter came within 30 feet of the aircraft it was intercepting, prompting a formal diplomatic protest by the United States to the MFA. According to an American official, the MFA seemed to know little about what China's military was doing, and suggested the pilots or the military units they were attached to were acting on their own. It would appear from such reporting that the April 1 collision was an accident waiting to happen. Furthermore, Lieutenant Shane Osborn, commander of the EP-3, would have been warned about the increasingly aggressive Chinese tactics before commencing his patrol.

In the morning of April 1 (March 31 in the United States), Lieutenant Osborn's EP-3 departed Kadena Air Force Base, Okinawa, on what would normally be a 9-hour patrol.[38] Two PLAN F-8-IIs joined up with the EP-3 shortly before the collision occurred around 9:00 a.m. According to the U.S. investigation of the incident, as the EP-3 turned eastward onto a 070-degree course to return to base, Lieutenant Osborn observed the two Chinese aircraft about a mile distant. As the F-8s began their interception, the EP-3 was on autopilot flying straight and level with an airspeed of 185-190 knots. The lead F-8 piloted by Lieutenant Commander Wang Wei then made two passes before the collision occurred during the third. On the first, he brought his aircraft to within 10 feet of the EP-3's port side and saluted the flight station. On the second pass, he had removed his oxygen mask when he closed to five feet and gestured to the EP-3's crew. On what was to be the final pass, the F-8 closed too quickly and was below and slightly forward of the EP-3's port wing when the pilot raised his aircraft's nose to bleed off some airspeed. In doing so, the Chinese fighter collided with the EP-3's number one propeller just forward of the F-8's vertical stabilizer and was torn apart. The

Chinese aircraft then crashed into the South China Sea some 70 nautical miles southeast of Hainan Island. Lieutenant Commander Wang Wei bailed out and was not recovered. As the result of the damage it received, immediately after the collision the EP-3 started a hard turn to the left. The U.S. Navy's investigation report suggests this may be why the other F-8 pilot, Commander Zhao Yu, reported that the American aircraft had caused the collision.

Bringing the severely damaged EP-3 under control, Lieutenant Osborn made an emergency landing at the PLAN's Hainan Island Lingshui air base after transmitting some 10 to 15 "Mayday" calls. The transmissions were not answered or, if they were, the EP-3 could not hear them because of the noise generated by the air rushing into the damaged aircraft.[39] Upon landing, the EP-3 was surrounded by Chinese soldiers and its crew of 24 detained on the island.

China's later insistence that the EP-3 was at fault is almost certainly based on Commander Zhao Yu's account of the collision when he returned to Lingshui air base.[40] Indeed, he could have interpreted the EP-3's post-collision sharp left turn as the cause rather than the consequence of the collision, as the U.S. Navy's investigation report suggests. However, two further aspects of China's interpretation are more difficult to accept. First, that the EP-3 did not request permission to land at Lingshui air base. Second, Commander Zhao Yu's contention that that the F-8s were flying a parallel course 400 meters from the EP-3's port side.[41] It is possible that the EP-3 was not using the radio emergency frequency most probably used by the Chinese military. If this were the case, then the Lingshui tower likely did not hear Lieutenant Osborn's "Mayday" transmissions.[42] The contention that the Chinese interceptors were 400 meters distant from the EP-3 is not supportable. Lieutenant Commander Wang Wei's history as an aggressive, risk-taking pilot[43] supports Lieutenant Osborn's report that F-8's distance was no more than 10 feet.

Response.

China made its position clear: The American aircraft was responsible for the collision. Moreover, the EP-3 had entered China's airspace and landed at Lingshui airbase without requesting permission to do so.[44] Consequently, China had the right to search the airplane and conduct its own investigation of the incident. For its part, the United States must accept its responsibility and apologize to China. Beijing's insistence on an apology from the United States was to be the central issue for both sides. The United States could not accept China's interpretation of the collision and therefore would not apologize. Ambassador Prueher, a retired Admiral and former commander of U.S. Pacific forces, was an experienced naval aviator. He knew that China's representation of the accident described an impossible situation. If the Chinese fighter had been 400 meters distant when the EP-3 banked to the left, the slow moving American aircraft about the size of the Boeing 737 would have passed behind the F-8. Nevertheless, the EP-3 and its crew were being held by China on Hainan Island. The United States wanted the 24 crew members and the EP-3 promptly returned, but did not want the incident to degenerate into a hostage crisis. Nor, as both sides worked toward a mutually acceptable resolution to the incident, it seemed did China.

Proceeding toward the negotiations was far more complicated than one would anticipate.[45] First, there was a lack of communication between the Chinese government and the American embassy. It was not until some 12 hours after the collision that China's MFA responded to Ambassador Prueher's request for a meeting. Assistant Foreign Minister Zhou Wenzhong, who was to be Ambassador Prueher's interlocutor throughout the negotiation, presented China's position on the incident and informed the ambassador that the EP-3's crew was safe. The following 3 days were a critical period. Admiral Dennis Blair, commander of U.S. Pacific Command (PACOM), and President Bush both had harsh words to say about the lack of communication between the Chinese government and the U.S. Ambassador. Around noon on April 2 (about midnight in Beijing), President Bush told reporters he was "troubled by the lack of a timely Chinese response" to Ambassador Prueher's request for access to the crew and their aircraft. He stated U.S. priorities as "prompt and safe return of the crew and return of the aircraft without further damage or tampering." First, however, China should grant embassy staff immediate access to the crew. Failure to do so, the President stressed, "is inconsistent with standard diplomatic practice and with the expressed desire of both our countries for better relations." [46]

On April 3, Defense Attaché Brigadier General Neal Sealock (USA), together with staff from the American embassy and the Consulate General in Guangzhou, were finally allowed to meet with the crew. General Sealock validated the crew list and learned how they were being treated and the extent of the damage to their aircraft. Because Chinese officials were present, General Sealock did not ask for Lieutenant Osborn's assessment of the accident. The following days saw movement toward the opening of negotiations. In a variety of venues, President Bush and Secretary Powell expressed their regrets and sorrow over the death of the Chinese pilot. President Bush sent a personal letter of condolences to Wang Wei's widow after receiving a letter from her charging him with cowardice. Although expressing regret and sorrow, the United States was equally insistent there would be no apology. Fearing a hostage crisis was looming, in the afternoon of April 4, Ambassador Prueher held a brainstorming session with his staff. After the meeting, Ambassador Prueher sent the outline of a negotiating strategy to Secretary Powell. Powell accepted the strategy and sent a letter to Vice Premier Qian Qichen proposing a series of steps to resolve the incident.[47] As he departed on April 4 for a scheduled 14-day state visit to Latin America, President Jiang Zemin repeated China's demand for an apology, adding the suggestion that the United States take a step favorable to easing Sino-American relations.[48] This last comment suggested Jiang was looking to resolve what was beginning to look like an impasse.

In sharp contrast to the May 1999 violent demonstrations around the American embassy in Beijing and consular facilities across China following the unintentional embassy bombing, in the aftermath of EP-3 collision the Chinese government kept a tight leash on potential demonstrators. Security around the embassy was tight, with police detaining the few protestors who did show up.[49] This behavior, together with President Jiang's departure for Latin America and China's acceptance of Secretary Powell's approach to resolving the incident, suggested Beijing was willing to accept a less conspicuous form of "apology" from the United States. No doubt, the expressions of regret and sorrow by President Bush and Secretary Powell had assisted in creating this atmosphere.

With Powell's letter as a road map, on April 5 Ambassador Prueher and Assistant Foreign Minister Zhou agreed to a 5-step negotiating process. For step 1, Secretary Powell agreed to the Chinese government publishing the first paragraph of his letter to Vice Premier Qian in which he expressed his regret over the loss of the Chinese pilot. Step 2 would be an official U.S. statement expressing condolences over the loss of life in the accident and regret that the EP-3 had entered China's airspace. Steps 3, 4, and 5 would focus on the release of the crew, an MMCA meeting to discuss procedures for avoiding future accidents, and arrangements for the return of the EP-3.

Negotiations.

Negotiations for the crew's release extended from April 6 to 11. The Chinese government sought a letter apologizing for the loss of its pilot and for the EP-3's entering China's airspace and landing at Lingshui air base without permission. The United States sought a factually accurate letter that did not apologize for anything related to the collision nor impede the President's ability to conduct relations with China in the future.[50] Ambassador Prueher and Assistant Foreign Minister Zhou usually met twice daily to negotiate the precise language of the letter. While the negotiations were underway, General Sealock met with the EP-3 crew on five separate occasions, beginning with a second visit on April 6.

By April 8, the United States was signaling that without some progress in the negotiations Sino-American relations would be damaged. During a CBS "Face the Nation" interview, Secretary Powell stated that "serious damage is now starting to be done" with congressional delegations canceling scheduled visits to China and businessmen wondering whether they should cancel theirs.[51] The next day, President Bush made the same point with reporters at the close of a cabinet meeting. He stressed that although effective diplomacy takes time, "Every day that goes by increases the potential that our relations with China could be damaged." The President concluded with the observation, "It is now time for our troops to come home so that our relationship does not become damaged."[52] The criticism President Bush was receiving from the conservative members of congress and media commentators such as William Kristol,[53] served only to increase pressure on the administration to reach a solution by taking a harder line with China. It is probable that China's embassy was reporting these pressures to Beijing.

Resolution.

What influence the public statements of President Bush and Secretary Powell had on China's assessments of the negotiations is not known. Nevertheless, by April 9 (April 8 in Washington), the U.S. embassy had the impression that resolution was near.[54] Despite the intense negotiations, however, they did not know when China would be satisfied with the text of the letter Ambassador Prueher was preparing for Foreign Minister Tang Jiaxuan. When the embassy learned on April 11 that Tang would meet Prueher later that day, it also appeared China was backing away from the negotiated agreements. The Chinese government wanted a signed letter before Ambassador Prueher met with the Foreign Minister, and would not say when the crew would be free to go. Both of these provisions violated agreements Ambassador Prueher had made with Assistant Foreign Minister Zhou. Finally, China wanted a U.S. official to sign a "memorandum of transfer" the Chinese had prepared. The "memorandum of transfer" was problematic for two reasons. It included language assigning responsibility for the collision to the United States—language Ambassador Prueher had specifically rejected. Additionally, some of the crew members names were spelled incorrectly.

Interpreting these new provisions as a Chinese effort to force a resolution without the negotiated compromises, Ambassador Prueher decided to be firm. Embassy personnel informed the Chinese government there would be no pre-signed letter. The ambassador would provide an unsigned letter with a cover note saying he would present a signed copy when he met with the Foreign Minister. On the matter of crew release, Ambassador Prueher was equally firm. The negotiated draft letter called for the crew to be released "as soon as possible." This meant immediately after the Foreign Minister accepted the signed letter. Ambassador Prueher expected Foreign Minister Tang to use just these words at the meeting. The ambassador also wanted assurances that China would provide flight clearance for the aircraft that would fly from Guam to Hainan to pick up the crew. Finally, although he saw no need for the "memorandum of transfer," Ambassador Prueher would agree to it if the names of the crew were spelled correctly, and the language assigning responsibility for the collision removed. The only language acceptable was that employed in the letter already negotiated. If China did not provide assurances on the crew's release and flight clearance for the American aircraft, there could be no meeting that day. When delivering these requirement, the embassy staff advised the MFA that if members of the Chinese media were at the meeting, or there was anything else that could potentially embarrass Ambassador Prueher, he would leave without handing his letter to the Foreign Minister.[55]

The meeting that day went as the U.S. embassy hoped. When Ambassador Prueher presented his letter, Foreign Minister Tang informed him the aircrew was free to leave. MFA staff provided flight clearance data for the chartered Continental aircraft that was to arrive at Haikou Airport at 6:00 a.m. the following day to pick up the EP-3 crew, refuel, and depart for Honolulu. With the crew's release on April 12, negotiations then turned to recovering the damaged EP-3 and arranging a special MMCA meeting focused on avoiding future aerial collisions.

At this point, U.S. lead in the negotiations passed from the Department of State to the Department of Defense (DoD).[56] Chinese negotiations continued to be an MFA responsibility. Both teams were at a lower official level than that used for negotiating the crew release. The U.S. delegation was led by Peter Verga, Deputy Under Secretary of Defense for Policy Support. The Chinese team was headed by Lu Shumin, a deputy to Assistant Foreign Minister Zhou Wenzhong. Mr. Verga brought two proposals to the meetings scheduled for April 18-19. First, that arrangements be made for the MMCA session devoted to avoiding future accidents. The second proposal was that the United States send an assessment team to determine whether the EP-3 could be repaired and flown out. If it could not be flown out, the aircraft should be moved by barge or disassembled and transported by some other method.

At the opening meeting, the Chinese team persisted in repeating China's position on the collision with no reference to the EP-3's return. The U.S. team was so frustrated by this that Mr. Verga threatened to walk out. Ambassador Prueher then arranged a meeting with Assistant Foreign Minister Zhou that resulted in the Chinese agreeing to discuss the EP-3's return the following day. When these initial discussions ended, the Chinese had agreed to study Verga's two proposals. The U.S. team then returned to the United States, and the American embassy assumed responsibility for negotiating the EP-3's recovery. On April 29, China agreed to a U.S. inspection team determining how the EP-3 should be recovered. China, however, would not permit the United States to repair the EP-3 so that it could be flown out. The United States also agreed to pay the cost of recovery.[57] The final agreement a month later was that the EP-3 would be dismantled and returned by cargo aircraft.[58] On July 3, the last components were loaded into a Russian-chartered An-124 cargo plane and the disassembled EP-3 began its final journey back to the United States.

With the return of the EP-3, acrimonious negotiations began over compensation for the crew's 11-day detention on Hainan and the cost of disassembling the EP-3. China presented a bill for $1 million, which the U.S. rejected as exorbitant.[59] About a month later, the United States said it would pay $34,567.00 as compensation, which China rejected as unacceptable. DoD insisted the amount was non-negotiable.[60]

More progress was made on the agreement that the United States and China would hold a special meeting of the MMCA to discuss ways of avoiding future accidents. On September 14-15, Chinese and American military representatives met on Guam where they discussed international law principles and procedures for ships and aircraft to follow when operating near each other. Both delegations agreed that the MMCA was the appropriate venue for assuring that air and maritime incidents be minimized in the future.[61]

Thus, although the compensation issue was yet unsettled, shortly after the 9/11 tragedy that was to dramatically change the environment of Sino-American relations, commitments made in the negotiations over the April 1 collision were completed. These negotiations, although frequently contentious, followed the pattern set by Secretary Powell's letter to Vice Premier Qian Qichen 5 months earlier.

Decisionmaking Under Stress.

Jiang Zemin faced much the same political dilemma he had to confront after China's Belgrade embassy was bombed. A Chinese naval officer had been killed in a collision with a U.S. military aircraft conducting hostile surveillance of China. What is more, the U.S. aircraft had penetrated China's airspace and landed at a Chinese military airfield without receiving permission to do so. In China's perception, the United States that Beijing had condemned for its "hegemonism" and identified as the principal threat to China's security in the defense white paper issued just 6 months earlier, was clearly at fault. Jiang Zemin had again to demonstrate that he was "tough" on the United States without severing the pragmatic relationship that served China's interests. He could not be seen as weak in confronting yet another example of American hegemonism and arrogance. As the days passed, the Chinese leadership may have suspected that Commander Zhao Yu's explanation of the collision was faulty. By then, however, the die was cast. China's position on responsibility for the collision and demand for an apology was established.

The United States was in a different situation. In such an accident, the United States understood international law and practice as requiring prompt return of the crew and their aircraft. As details of the collision became known from General Sealock's discussions with the EP-3 crew, the United States became even more irate with China's position. Nonetheless, with the crew and aircraft in China's control, Washington had to avoid placing so much pressure on Beijing that it believed only continued detainment of the crew would force any kind of U.S. concession. Should this happen, then the United States would face a hostage crisis.

Although the United States and China had distinctly different perceptions of what the collision implied, both had good reason to work out a diplomatic compromise. Thus, avoiding a "hostage crisis" while achieving their objectives seems to have been the negotiating strategy followed by both China and the United States. The United States wanted the quickest resolution possible. Jiang Zemin and the Chinese leadership appeared to believe they would gain the most by prolonging the resolution process. Jiang would be seen as tough on the United States, and China would be seen as refusing to bow under American pressure. As in the Belgrade embassy bombing incident, U.S. interest in a quick solution served Jiang Zemin well. Furthermore, controlling the U.S. crew and aircraft, but especially the crew, provided China an advantage in the negotiations. The United States may well have had international law and practice on its side, but China had physical control of U.S. military personnel and their aircraft. China could therefore end the negotiation process, when it concluded two political objectives had been achieved. First, when Jiang Zemin believed he had satisfied his political objective of being seen as resolute in dealing with the United States. Second, when sufficient time had passed and the incident received enough global media attention for China to be perceived as being strong enough to squeeze concessions out of the United States.

From Beijing's perspective, this strategy could be seen as successful. First, President Bush and Secretary Powell expressed their regret and sorrow for the consequences of the collision. Then Ambassador Prueher's tightly negotiated letter declared that the United States was "very sorry" for the loss suffered by the family of Wang Wei, and was "very sorry" that the EP-3 entered China's airspace without verbal permission.[62] The Chinese press, as one would expect, declared these expressions to be the apology Beijing had demanded and therefore the incident concluded with diplomatic victory for China.[63] The United States, of course, had a different view. The United States had accepted no responsibility for the collision and had not formally apologized. An anonymous White House "senior official who was deeply involved in the negotiations" asserted that China's decision to end the confrontation came only after President Bush and Secretary Powell warned that the entire relationship was at risk.[64] Nonetheless, China's negotiating tactics became stiffer in the days immediately after President Bush and Secretary Powell expressed their discontent with the lack of progress being made by the negotiations.

Where China and the United States meshed was in their efforts minimizing the possibility that the collision and its consequences become a hostage crisis. As President Bush sought to ease American public opinion and criticism from political conservatives by stressing that diplomacy takes time, so China prevented demonstrations, spontaneous or otherwise, around the American embassy's compound.

Beijing's insistence on detaining the EP-3's crew for 11 days, however, did little to ease the Bush administration's suspicions of China as a strategic competitor. Nonetheless, testifying before the Congress 2 months after the EP-3's crew had been released, Assistant Secretary of State for East Asian and Pacific Affairs James Kelly sought to place the U.S. posture in clearer focus. In his testimony, although carefully stating that China was not an enemy and listing the areas where China and the United States had complementary interests, Secretary Kelly was quite frank in his assessment. He observed,

> Recent events have called into question where we stand in our relationship with China and where we want to go. They have highlighted the importance of not allowing our relationship to be damaged by miscommunication, mistrust and misunderstanding about our respective intentions and objectives.

Despite this admonition, much of Secretary Kelly's testimony highlighted the most positive aspects of Sino-American relations. The United States continued to support China's entry into WTO and its hosting of the coming APEC summit in Shanghai. Perhaps most telling, Secretary Kelly underscored President Bush's intent to attend the APEC summit and go to Beijing. In short, the United States wanted to put the EP-3 incident behind it and move ahead.[65] China clearly wanted to do the same. Indicative of China's intent was its response to the renewed reconnaissance patrols that had been suspended during negotiations over the EP-3 crew's release. Speaking to reporters flying with him on the way to Australia after a 1-day stop in Beijing, Secretary Powell stated that Chinese pilots had stopped using the aggressive tactics that had led to the F-8's collision with the EP-3. He viewed this change as but one of many signs China wanted to improve relations with the United States.[66]

Restoring the military relationship severed by the United States in response to China's detention of the EP-3 crew was the slowest of all moves toward normalization. Despite the *rapprochement* that occurred in other realms of Sino-American relations following the 9/11 tragedy, military relations remained almost on hold. It was not until December 2002 that the Defense Consultation Talks were restored with General Xiong Guangkai's meeting with Douglas Feith, the Under Secretary of Defense for Policy.[67] In all other respects, Sino-American relations moved swiftly toward normal beginning in the fall of 2001 when the United States reordered its security priorities to focus on global terrorism.

CONCLUSIONS AND SPECULATIONS

In the best of times, China views the United States with apprehension. In the worst of times, this apprehension degenerates into hostility. Beijing reluctantly accepts that China's pragmatic interests require it to avoid sustained confrontation with the United States and to cooperate where possible. This reluctant recognition of American power and China's need for U.S. cooperation enhances Beijing's sensitivity to what it perceives as American arrogance and "power politics." This core context of apprehension and hostility joined with Beijing's recognition that China cannot afford and does not have the power for sustained confrontation with the United States was at the root of Beijing's responses to these incidents.

Despite the dramatic differences between them, a pattern emerges common to both cases. In each, Beijing applied an asymmetric strategy designed to exploit the U.S. quest for a quick resolution to the incidents. Beijing evidently believed that Washington's desire to end the diplomatic impasse stemming from these incidents could be manipulated into granting China the initiative in the negotiations. Beijing thus prolonged the negotiations. It did so in order to achieve quite specific political objectives. First, to highlight Chinese patriotism and nationalism by demonstrating to the world that regardless of the power differences between them, China could not be cowed by U.S. might. In Beijing's view, the longer it took to resolve the incidents, the stronger China would appear to be. Second, by prolonging resolution of the incidents China sought to demonstrate that the United States was the supplicant in the negotiation process. Third, extending negotiations was seen as way of publicly extracting as many concessions from the United States as possible. Of particular importance were public expressions of regret and sorrow by Presidents Clinton and Bush. Fourth, the strategy was to prolong the negotiations but not to the point where the protraction would cause severe damage to the bilateral relationship.

It must be noted that this strategy was made possible by two specific conditions. The United States immediately announced that the bombing of China's embassy was a tragic error. In the EP-3 incident, China had physical control of the crew and the aircraft. Without such favorable conditions, employing an effective asymmetric negotiating strategy would be far more difficult.

Although these incidents were military in origin, the role China's military leaders played in the diplomacy resolving them is completely opaque. It can be safely assumed that the PLA's leadership had a voice in determining what China's responses should be, but there was no overt military presence in either set of negotiations. It was evident that China was keeping the negotiating process in the hands of professional diplomats. Interestingly, the negotiating strategy pursued by China mirrored a core doctrinal principal held by the PLA: gain and hold the initiative.

ENDNOTES - CHAPTER 6

1. This discussion draws extensively from Joseph Fewsmith, *China Since Tiananmen: The Politics of Transition*, NY: Cambridge University Press, pp. 208-224; and Robert L. Suettinger, *Beyond Tiananmen: The Politics of U.S.-China Relations 1989-2000*, Washington, DC: Brookings Institution Press, 2003, pp. 328-407.

2. See David M. Finkelstein, *China Reconsiders Its National Security Strategy: "The Great Peace and Development Debate of 1999,"* Alexandria, VA: The CNA Corporation, December 2000, for a full analysis of this issue.

3. Fewsmith, *China Since Tiananmen*, p. 209; and Finkelstein, *China Reconsiders its National Security*, p. 17.

4. The following discussion draws extensively from Suettinger, *Beyond Tiananmen*, pp. 358-369.

5. The House of Representatives Select Committee on U.S. National Security and Military/Commercial Concerns with the People's Republic of China chaired by Representative Christopher B. Cox.

6. The following discussion is taken from Zong Hairen, *Zhu Rongji zai 1999* (Zhu Rongji in 1999), "Visit to the United States," translated in *Chinese Law & Government*, Vol. 5, No. 1, January-February, 2002, pp. 36-52.

7. Zong Hairen, *Zhu Rongji in 1999*, p. 40; Suettinger, *Beyond Tiananmen*, p. 365; and Fewsmith, *China Since Tiananmen*, p. 209.

8. Fewsmith, *China Since Tiananmen*, p. 211.

9. Zong Hairen, *Zhu Rongji in 1999*, "The Bombing of China's Embassy in Yugoslavia," p. 75.

10. *Ibid.*, pp. 74-75.

11. Observers on the scene remarked on this genuine anger, as did James Mulvenon in a personal communication on May 21, 1999.

12. Suettinger, *Beyond Tiananmen*, p. 371.

13. *Ibid.*, p. 372.

14. Beijing, Associated Press, 11:33 p.m. EDT.

15. Suettinger, *Beyond Tiananmen*, p. 373.

16. *Ibid.*, p. 372.

17. Beijing, *Xinhua* Domestic Service, May 13, 1999.

18. Fewsmith, *China Since Tiananmen*, p. 212; and Suettinger, *Beyond Tiananmen*, p. 374.

19. UN Security Council Press Release SC/6675, May 14, 1999.

20. Beijing, *Xinhua*, June 12, 1999.

21. Oral Presentation by Under Secretary of State Thomas Pickering on June 17 to the Chinese Government Regarding the Accidental Bombing of the PRC Embassy in Belgrade, U.S. Department of State, Office of the Spokesman, released July 6, 1999. *Xinhua* says the briefing was given on June 16.

22. Joseph Fewsmith, "The Impact of the Kosovo Conflict on China's Political Leaders and Prospects for WTO Accession," NBR Publications: NBR Briefing, July 1999.

23. Beijing, *Xinhua*, June 17, 1999.

24. Suettinger, *Beyond Tiananmen*, p. 376.

25. Bonnie S. Glaser, "Sino-American Relations Challenged by New Crises," *Comparative Connections*, Vol. 1, No. 1, July 1999, p. 7.

26. Beijing, CNN, July 30, 1999.

27. Transcript, "Responses to Questions Submitted by Deutsche Welle. Lee Teng-hui, President, Republic of China," pp. 2-3. Provided by the Taipei Economic and Cultural Representatives Office in the United States, July 14, 1999.

28. See Suettinger, *Beyond Tiananmen*, pp. 385-388, for an insightful assessment of this period

29. Transcript, State Department Legal Adviser David Andrews at Conclusion of Negotiations, Beijing, China, December 16, 1999.

30. *The One China Principle and the Taiwan Issue*, Beijing: Taiwan Affairs Office and the Information Office of the State Council, February 21, 2000.

31. *China's National Defense 2000*, Beijing: Information Office of the State Council, October 16, 2000.

32. See, for example, Governor Bush's speech at the Ronald Reagan Public Library, "A Distinctly American Internationalism," November 19, 1999; and Condoleezza Rice, "Promoting the National Interest," *Foreign Affairs*, Vol. 7, No. 1, January-February 2000, p. 56.

33. *China's National Defense 2000*, p. 3.

34. "Confirmation hearing by Colin L. Powell," Department of State, Washington, DC, January 17, 2001.

35. John Pomfret, "Jiang has caution for the U.S. China's Leader Says Taiwan Arms Deal would Spur Buildup," *Washington Post*, March 24, 2001.

36. The following discussion is drawn from Thomas E. Ricks, "Anger Over Flights Grew in the Past Year. Proximity Riled China; U.S. Cited Interceptors," *Washington Post*, April 7, 2001.

37. See Article II, "Agreement between the Department of Defense of the United States of America and the Ministry of National Defense of the People's Republic of China on Establishing a Consultation Mechanism to Strengthen Military Maritime Safety," signed in Beijing, January 19, 1998.

38. This description of the collision is drawn from "Statement of Facts Regarding Chinese Intercept Aircraft's Collision with U.S. EP-3, April 16, 2001, 1630," obtained by *Judicial Watch* through the Freedom Of Information Act, FOIA, "Documents Related to the U.S. Navy EP-3's Collision and Emergency Landing on Hainan Island" (n.d.), released by the Commander in Chief, U.S. Pacific Fleet, Pearl Harbor, Hawaii.

39. Transcript of a telephone call between Secretary of Defense Donald H. Rumsfeld and Lieutenant Shane Osborn provided by the Armed Forces Press Service, Washington, April 13, 2001.

40. See James Mulvenon, "Civil-Military Relations and the EP-3 Crisis: A Content Analysis," *China Leadership Monitor*, No. 1, Winter 2002, for a full discussion of this possibility.

41. Interview with Commander Zhao Yu, Beijing, *Xinhua*, April 6, 2001.

42. Personal communication with Rear Admiral Eric McVadon, USN, ret., April 15, 2001.

43. Steven Lee Myers with Christopher Drew "Chinese pilot reveled in risk, Pentagon says," The *New York Times*, April 6, 2001.

44. "Chinese Foreign Ministry Spokesman Gives Full Account of Air Collision," Beijing, *Xinhua*, April 4, 2001.

45. The experience of Ambassador Joseph Prueher and the United States embassy staff is drawn from John Keefe, *Anatomy of the EP-3 Incident, April 2001*, Alexandria, VA. CNA Project Asia, January 2002. Mr. Keefe was special assistant to Ambassador Prueher at the time of the EP-3 incident.

46. "Statement by the President on the American plane and crew," The White House, Office of the Press Secretary, 11:38 EDT, April 2, 2001.

47. Keefe, *Anatomy of the EP-3 Incident*, p. 7.

48. "U.S., China Soften Words, Seek Way Out," *Seattle Times*, April 5, 2001.

49. Kenneth T. Walsh, "The China Conundrum," *U.S. News & World Report*, April 16, 2001; and Keefe, *Anatomy of the EP-3 Incident*, p. 13.

50. Keefe, *Anatomy if the EP-3 Incident*, p. 8.

51. Secretary Colin L. Powell, Interview on CBS "Face the Nation" by Bob Schieffer of CBS and Tom Friedman of *The New York Times*, U.S. Department of State, April 8, 2001.

52. "President Bush Warns Risk of Damage to U.S.-China Relations Increases Daily," U.S. Department of State, Office of International Information Programs, April 9, 2001.

53. William Kristol, intervied by Mara Liasson, "All Things Considered," National Public Radio, April 10, 2001.

54. Keefe, *Anatomy of the EP-3 Incident*, p. 10.

55. *Ibid.*, p. 11.

56. This discussion draws heavily from Keefe, *Anatomy of the EP-3 Incident*, pp. 12-13.

57. Beijing, *Xinhua*, April 29, 2001.

58. Beijing, *Zhongguo Xinwen She*, May 24, 2001.

59. Alan Sipress and Thomas E. Ricks, "China Bills U.S. Over Collision. $1 Million in Surveillance Plane Dispute Termed 'Exaggerated'," *The Washington Post*, July 7, 2001.

60. Elisabeth Rosenthal, "China Spurns Spy Plane Offer as Inadequate," *The New York Times*, August 13, 2001.

61. "U.S. Concludes Military Talks with China," U.S. Department of Defense News Release, No. 432-01, September 15, 2001.

62. Ambassador Prueher's letter can be found in Keefe, *Anatomy of the EP-3 Incident*, Appendix A.

63. See, for example, He Chong, "China wins diplomatic victory in forcing U.S. to apologize," Hong Kong, *Zhongguo Tong She,* a PRC-owned press agency, April 12, 2001; and Yu Shuyang and Hu Siyuan, "The United States Peremptoriness is Rebuffed in the End," Beijing, *Zhongguo Guofang Bao*, April 13, 2001.

64. David E. Sanger and Steven Lee Myers, "Delicate Diplomatic Dance Ends Bush's First Crisis," *The New York Times*, April 12, 2001.

65. U.S. Department of State, "United States Policy in East Asia and the Pacific: Challenges and Priorities." Testimony of James A. Kelly, Assistant Secretary of State for East Asian and Pacific Affairs before the Subcommittee on East Asia and the Pacific of the House Committee on International Relations, June 12, 2001.

66. Philip P. Pan, "Powell says China Eases Pursuit of U.S. Jets," *The Washington Post*, July 30, 2001.

67. James Dao, "U.S. and China Resume High-Level Military Talks," *The New York Times*, December 10, 2002.

CHAPTER 7

DECISIONMAKING IN TRIPLICATE: CHINA AND THE THREE IRAQI WARS

Yitzhak Shichor

INTRODUCTION

Over the past quarter of a century, Iraq has been the progenitor of three international confrontations: the Iran-Iraq War (1980-88, hereafter the first Gulf War); Saddam Hussein's invasion of Kuwait that triggered a U.S.-led offensive under United Nations (UN) auspices (1990-91, hereafter the second Gulf War); and the Iraq War, launched by a U.S.-led coalition in the spring of 2003 (hereafter the third Gulf War). Though not identical, these conflicts have taken place in the same region, with similar players, reflecting similar circumstances and around similar issues, thereby providing outstanding case studies for a comparative analysis.[1] Iraq, the culprit in all three, first attacked Iran, then invaded Kuwait, and finally refused to expose the weapons of mass destruction it had supposedly accumulated. While the first conflict had remained basically regional and bilateral, the other two involved a coalition led by an American Republican administration (headed by a president of the same family and same officials). Both offensives aimed at removing a leader considered a regional (and some would say a global) threat. What was China's role in the three conflicts?

To be sure, the People's Republic of China (PRC) was not directly involved in any of them. Nevertheless, Beijing could by no means overlook these conflicts. It was forced to react not only as a leading member of the international community, but also because some of its regional and global interests were at stake. Willingly or not, the PRC was implicated in these conflicts indirectly, both actively and passively, and, furthermore, could have affected their outcome. Still, in all these cases, Beijing's decisionmaking was intended less to avert the confrontations, slow them, or stop them, and more to minimize its own losses and maximize its own gains. Determined by its international, regional, and domestic conditions,

Beijing's reaction to the three conflicts respectively could have, and eventually has, affected some of its regional, domestic, and global interests. Ultimately, and given the differences among the three conflicts, Beijing's response has apparently been almost identical and very consistent—less by choice and more by necessity. In this respect, China's handling of the Iraqi wars can be regarded as decisionmaking in triplicate.

Basically, the same pattern has been repeated in all three crises. While attempting to please all parties concerned, the Chinese have always tried to dissociate themselves from any direct involvement in the conflict. As a rule, Beijing has always called for a peaceful settlement of the conflict, preferably by the parties concerned or their peer (Arab, Afro-Asian, Third World) governments. UN Security Council intervention was less acceptable, not only because of Beijing's own negative historical experience but primarily because of the Chinese unwillingness to becoming involved in the conflict resolution process which, by necessity, would mean alienating one of the parties. Also as a rule, China has firmly opposed the use of force to settle conflicts, both under and definitely outside UN auspices. This policy, again, has not been based only on principles but also on expediency. Post-Mao China needed international stability to proceed with its economic development and modernization, which could be disrupted by a military confrontation. Still, military action has not been without its benefits. For one thing, it provided China with an opportunity to sell arms, often to all sides. For another, a point that will be elaborated later on, it provided Beijing with a golden opportunity to become exposed to more advanced military technologies and invaluable intelligence—impossible to get otherwise—without any risk, either direct or indirect. China's double standards and subterfuge have reached their climax in response to external conflicts, far away from its borders. A different set of rules exists for dealing with conflicts nearby.

If uncovering the dynamics of decisionmaking by any government is difficult, it is next to impossible in the case of the Chinese. For one thing, China's decisionmaking processes are far less transparent compared to other countries' and cases of dealing with external conflicts are hardly touched by existing studies.[2] For another, a certain distance in time is needed to provide for declassified and archival documents, for personal reminiscences and for interviews. All are now available for the Korean War, and indeed shed more light on China's decisionmaking *then*, but are not available for the first Gulf war, let alone for the second and third. Under these circumstances, the only way to study Beijing's decisionmaking in the three Gulf conflicts is by reaching conclusions on the basis of a reconstruction of events; an interpretation of responses; an evaluation of the consequences; an understanding of the Chinese mind and way of thinking—and a good deal of intuition. This "method" is adopted hereby.

THE FIRST GULF WAR: ELEMENTARY DECISIONMAKING

On September 22, 1980, Iraqi forces invaded Iran in an attempt to seize control of Khuzestan Province, including a large part of Iran's oil wealth. Failing to reach a swift military victory, Iraqi forces were bogged down in an attrition war that lasted until 1988. The eruption of the war between two countries considered friendly to them presented the Chinese with a dilemma. In 1958 Baghdad had become the fourth Arab government to establish full diplomatic relations with the PRC.[3] Despite occasional tension—caused by the persecution of Iraqi communists in the late 1950s, Iraq's growing dependence on the Soviet Union, and China's anxiety about Iraq's destabilizing role in the Persian Gulf—relations between the two countries developed smoothly especially since the mid-1970s.

By that time, relations between China and Iran had gathered momentum. Regarded as an outpost of "U.S. imperialism" in the Middle East and as an opponent of the PRC, Iran finally recognized Beijing only in August 1971, precisely when Sino-U.S. relations had begun to improve. Under these circumstances, China now perceived Iran's association with Washington and firm opposition to Moscow as an asset. This Chinese perception was perhaps best illustrated by a series of high-level visits crowned by Chairman Hua Guofeng's arrival in Teheran in August 1978. This was the first ever visit by China's supreme leader to any Middle Eastern country. Yet this Sino-Iranian honeymoon was short-lived. Soon after Hua's visit the U.S.-backed Shah's *ancien régime* disintegrated, paving the ground for a radical Islamic government. Beijing's brief romance with the Shah suddenly became a liability. However, despite its distaste for Islamic radicalism, Beijing had to act swiftly to restore its good relations with Teheran. By the time of the Iraqi invasion, China had already taken steps in this direction. The war could be used to further consolidate this process, essential not only in bilateral or trilateral terms but also in terms of China's domestic and international situation.

Domestically, China's post-Mao leaders had just launched far-reaching reforms that, definitely in a retrospective view, have been nothing less than dramatic. Despite early uncertainties they represented firm commitments to modernization and development, overturning Mao's legacy of continuous revolution. Gradually dismissing the theory of an imminent third world war, the Chinese perceived the international arena in more benevolent terms, conducive to their own agenda of accelerated economic growth. To be sure, the international landscape had also changed. For the first time since the PRC establishment, Sino-American official relations had been formed that allowed China not only to better integrate in the world community, but also to confidently oppose the Soviet Union. Regarded since the late 1960s as the most immediate threat to China's security, Moscow was facing an unprecedented Beijing-Washington partnership in Asia (where its troops were bogged down in Afghanistan), and the Middle East. These regional, domestic, and international situations had determined China's response to the first Gulf war.

One day after the outbreak of the war Prime Minister Zhao Ziyang expressed concern about the armed conflict and set out Beijing's three main official objectives—that the conflict would be settled peacefully through negotiations; that it would remain free of superpower intervention; and that it would not deteriorate.[4] In words, these objectives appeared to be consistent, but not in deeds. As the only permanent member of the UN Security Council with good relations with both Iraq and Iran, the PRC could have led the way in mediating between them toward reaching a peaceful settlement to the war, according to its official stand, but did not. Beijing's reluctance to offer its mediating services originated in its unwillingness to take sides, primarily in regions beyond its immediate interests and comprehension, and in its overall suspicion of the UN. Nine years after its admission to the UN, Beijing was still feeling, and behaving, like an outsider. It was only toward the late 1980s that the Chinese began to realize their UN power—and use it, though cautiously.[5] It was only *following* the adoption of UN Security Council Resolution 598 on July 20, 1987, calling for Iraq and Iran to cease fire and start peaceful settlement negotiations, that Beijing finally sent its deputy foreign minister Qi Huaiyuan as a special envoy to Teheran (August 25, 1987). Within 1 year, on August 20, 1988, the Iran-Iraq war ended—not necessarily thanks to China or in line with its interests.

Allegedly, the Chinese did not want the war to deteriorate or extend. This could have undermined their economic growth in general and interests in the Gulf in particular. To be sure, throughout the conflict, the Chinese consistently criticized the superpowers for sustaining the war, thereby causing human suffering and economic devastation. In fact, the war was not as detrimental to China's economic (and military) interests as we might have assumed. By 1987, just before the war was over, Iraq had become China's number one market for labor service cooperation (labor export), valued at US$657.67 million, or nearly 70 percent of the total! Iraq had also become Beijing's number one market for contracted projects (construction services), valued at US$670.04 million, or over 18 percent of the total. Altogether, China's turnover for these activities with Iraq alone reached US$1.328 billion, or nearly 30 percent of the total. These figures are amazing, given the fact that these activities began in 1979 and that Iraq was at war for most of the time. While the war hardly affected China's economic interests in Iraq, its conclusion could undoubtedly benefit the Chinese even more. Yet, notwithstanding its statements, Beijing—very much like Washington and Moscow, perhaps even more—must have welcomed the continued Iraqi-Iranian deadlock and, moreover, directly and soberly contributed to its persistence by supplying weapons to both sides.

It is possible that Beijing's decision to supply arms to Iraq and Iran had already been made before the war. Washington's termination of its military supplies to Iran after the Shah's downfall and the slowdown in Moscow's arms exports to Iraq in the late 1970s had provided China with an opportunity. This opportunity could become even more lucrative with the eruption of the war and the anticipated erosion of weapons on both sides. Beijing could not stand the temptation. Of the four options of selling only to Iraq, only to Iran, to neither, or to both, Beijing opted for the last, although by any standard it contradicted its rhetorical insistence on a peaceful and speedy end of the conflict. Regular Chinese arms supplies to Iraq and Iran began in 1981-82, covering more or less the same weapons: hundreds of fighter planes, tanks, artillery pieces, and armored personnel carriers, and thousands of missiles of different kinds. In the 1980s Iraq and Iran became China's leading arms market, valued at US$7-7.5 billion, around 55 percent of all Chinese arms agreements and nearly 70 percent of Beijing's total arms deliveries.[6]

To be sure, Beijing consistently denied selling arms to Iraq and Iran during the war. To some extent, it was right. As early as 1987, I pointed out that Chinese arms to the two belligerent states had been sold indirectly.[7] Unclassified trade statistics published by the International Monetary Fund (IMF), as well as in Chinese customs and trade statistics revealed the incredible swell in Chinese exports to Jordan, a small and underdeveloped country of 2.4 million. The value of Chinese exports to Jordan jumped from practically nothing to about US$1.32 billion in 1982, to US$1.53 billion in 1983, and to US$1.26 billion in 1984. Overnight Jordan became China's fourth export market preceded by such economic giants as Hong Kong, Japan, and the United States and still outranking Singapore! There is no doubt that Jordan provided a clearinghouse, and partly also a channel, for the Chinese military supplies to Iraq (it is interesting that Jordan's import statistics by no means even come close to China's export statistics and reflect a huge gap). Other regional governments concerned about Teheran's territorial designs in the Gulf may have provided a channel for Chinese arms to Iraq (e.g., Egypt, Saudi Arabia, and Kuwait). At least some Chinese arms may have been channeled to Iran indirectly by Syria and perhaps Pakistan. Officially, all direct, and even indirect, arms sales to Iraq and Iran—both designated as "sensitive" countries—had been generally forbidden. Exceptions must have been cleared by the Ministry of Foreign Affairs and approved by the central leadership. China's use of proxies to supply Iraq and Iran has been confirmed much later. The flow of Chinese arms to Iraq and Iran, occasionally without official endorsement, was also the outcome of the fierce competition between the People's Liberation Army's (PLA) arms exports drive and that of the defense industrial establishment.[8] Firmly denying that China had sold arms to Iran and Iraq, PRC Ministry of Foreign Affairs' spokesman Ma Yuzhen pretended innocence: "The International arms market is very complicated . . . Therefore, we have no way of finding out how other countries procure their weapons from this market." Asked if China would take action to prevent Chinese arms from reaching Iran or Iraq through indirect channels, he declined to comment. In a similar vein, when interviewed by NBC, Premier Zhao Ziyang failed to concede that Iran had acquired Chinese arms—either directly or indirectly. Pleading inconceivable ignorance, he confided: "Up to now I still do not believe that the missiles Iran has are Chinese missiles . . . acquired through other channels. [However] if a country has the money and is willing to offer a high price, there will be no difficulty for this country to find channels in which it can acquire weapons."[9] Needless to mention, both Iraq and Iran were also supplied directly by the Chinese.

Year	Iraq	Iran	Jordan	Saudi Arabia	Kuwait	Egypt	Syria
1978	69	65	—	37	93	54	—
1979	135	37	—	68	136	69	—
1980	124	121	—	136	157	215	—
1981	200	149	—	168	146	230	—
1982	120	41	1,287	180	109	254	71
1983	35	267	1,516 (1,582)*	149	99	205	166
1984	55	155	1,235 (1,264)*	133	85	167	343
1985	128	84	984 (1,060)*	133	73	98	66
1986	151	50	1,031 (940)*	134	73	130	320
1987	74	94	1,341	247	94	125	382
1988	75	222	723	230	131	166	184

Source: International Monetary Fund, *Direction of Trade Statistics Yearbook 1983, 1990.*

Almanac of China's Foreign Economic Relations and Trade, 1984-1986.

Table 1: China's Export to Selected Middle Eastern Countries, 1978-88 (In Million U.S. Dollars).

Apparently, Beijing provided these arms supplies primarily in terms of the Iran-Iraq war as an attempt to prevent a radical change in the regional power balance that could deteriorate to chaos and instability. Intended also to bolster China's relations with the two countries, these supplies allegedly had been aimed at undermining and perhaps excluding the Soviet presence in the Persian Gulf. These sophisticated interpretations, however, conceal the possibility that the simple motivation for China's arms supplies to both Iraq and Iran had been economic, or to be more precise, military. In addition to the income earned from these arms sales—that was channeled, at least partly, to fuel China's own defense modernization—the Iran-Iraq war provided a testing ground for Chinese weapons under battlefield conditions. Coming shortly after China's poor performance in Vietnam, the Iran-Iraq war was a godsend. Chinese-made weapons could be tested against Iran's U.S.-made weapons and Iraq's Soviet-made weapons—simultaneously. Viewed in these economic and military perspectives, the early conclusion of the war was the last thing Beijing wanted. It was only in 1987, when Chinese-made weapons were used by Iraq and Iran to threaten, and eventually harm, U.S. interests and traffic in the Gulf, that the bilateral conflict was internationalized and efforts to settle the conflict were given a push, leading to the end of the war in 1988.

Though occasionally regarded as a sideshow and neglected by academic research, the first Gulf War was crucial in determining China's attitudes and responses to external conflicts. The Chinese response to the Iran-Iraq war can be termed "elementary" decisionmaking because the equation and the choice were rather simple. Only two countries were directly involved in the conflict, which the two superpowers and the rest of the international community, including the UN, basically ignored. Until 1987 there was no real threat of diplomatic, political, let alone military intervention that would have required Beijing to determine its policy beyond the rules that had been adopted in 1980. These called for making the most out of the crisis with as little involvement as possible and without antagonizing the parties concerned, whether regionally or globally. These rules were applied yet again in the second Gulf War when the decisionmaking equation and the potential choices became a little more complex.

THE SECOND GULF WAR:
INTERMEDIATE DECISIONMAKING

Although the Chinese must have already been aware of Saddam Hussein's aggressive predilections, his assault on Kuwait on August 2, 1990, caught them (and most others) by surprise.[10] Apparently, the second Gulf conflict began very much like the first, which had ended only 2 years earlier. By that time Beijing had already realized that not all the Third World problems could be blamed, in Mao's style, on Western imperialism. For many of these problems had originated in genuine conflicts and particularistic issues that had had nothing or little to do with U.S. imperialism or Soviet social-imperialism. Reached in the mid-1980s, one of China's most important strategic lessons of the Iran-Iraq War was that localized conflicts and protracted wars became the main threat to regional and international stability, rather than an imminent nuclear world war involving the two superpowers. At the beginning, the second Gulf conflict had promised to be a reincarnation of the Iran-Iraq War, but it soon turned to be something completely different, definitely from Beijing's perspective. Conforming so nicely to their internal and external agenda in strategic, military, and economic terms, the first Gulf conflict had paralyzed the Chinese into believing that the next regional war would be very much the same as the last. It is this belief that somewhat explains China's "profound shock" in view of the Western intervention, and even more so—military performance. Supposedly, both had been unexpected, not to mention that, when the conflict erupted, the PRC was not only unprepared but was, moreover, stuck in a triple quagmire, in the Middle East, at home, and internationally.

To begin with, Beijing used to have traditionally good though contradictory relations with both Iraq and Kuwait. In 1961 Kuwait, which had long been regarded by Iraq as an integral part of its own territory, was granted independence by the British, creating a dilemma for Beijing. Instant recognition of Kuwait's independence and liberation from British colonialism must have been the right thing to do. Instead, however, the Chinese chose the traditional midway by avoiding direct recognition of Kuwait, yet supporting its right to independence from the British. They soon realized their gross miscalculation. In 1963, only after more than 70 governments (among them Taiwan) already had recognized Kuwait, did the Chinese began negotiations that finally led to the overdue establishment of diplomatic relations between the two countries on March 22, 1971—nearly 10 years after Kuwait had become independent.[11]

Since then, Beijing's relations with Iraq and Kuwait developed smoothly, primarily in economic terms. As mentioned above, by 1991 Iraq had become China's primary market for labor export—over four times greater than the next market (Hong Kong)—and China's second most important market for contracted projects (after Hong Kong). Around 60 Chinese companies and some 4,000 workers, including military staff, were caught in Iraq throughout the war and could not be evacuated, unlike those in Kuwait. Kuwait had also been one of China's leading markets for labor export and contracted projects (at US$124.7 million and US$363.66 million, respectively). When these new activities began in 1979, Iraq and Kuwait had been China's nearly exclusive markets for labor export and construction services. Kuwait had also provided China with loans and foreign direct investment, primarily for its energy economy.[12]

Furthermore, as mentioned above, by the time Saddam invaded Kuwait, Iraq had been China's leading arms customer. By 1990 the Iraqi share in China's arms deliveries had reached nearly one-third (about US$4.2 billion)—more than any other country. They included six B-6 (H-6) bombers, 128 C-601 anti-ship missiles, 72 HY-2 (*Haiying*, also known as Silkworm) missiles, 700 Type-59 MBTs, 600 Type-69 main battle tanks (MBTs), 650 Type 531 armored personnel carriers (APCs), and 720 Type 59/1 130mm towed guns.[13] At the time of the Iraqi invasion of Kuwait Chinese-made armor accounted for nearly one-third of the Iraqi armored forces. There is also some evidence of Chinese attempts to supply Iraq with chemicals that could be used for the production of missile fuel and perhaps also of nuclear weapons.[14] Still, Iraq and Kuwait were only one part of China's crisscross interests in the Persian Gulf.

After many years of unwillingness, and following the acquisition of Chinese-made DF-3 intermediate range ballistic missiles (IRBMs), Saudi Arabia finally decided to establish full diplomatic relations with the PRC on July 21, 1990, some 12 days before Saddam Hussein's invasion of Kuwait. For China, and especially in a retrospective view, this was a major coup that also included a serious blow to Taiwan. Riyadh, Iraq's rival and Kuwait's ally, could by no means be overlooked by Beijing—and the same goes for Iran that the PRC had systematically cultivated since the downfall of the Shah in 1979, primarily with arms supplies. Also, since the launch of its post-Mao reform, Beijing managed to establish diplomatic relations with all Persian Gulf countries, including Oman (May 25, 1978), the United Arab Emirates (November 1, 1984), Qatar (July 9, 1988) and Bahrain (on April 18, 1989). In sum, China had a good deal at stake in the Persian Gulf, not only economically and militarily, but also politically: all these countries, without exception, also chose to stand by the PRC during and after the Tiananmen confrontation, a stand China had to take into account.

The Tiananmen confrontation, its internal and even more so international outcome, governed China's role in the second Gulf conflict. This was the beginning of the Sino-American mutual disillusionment following a decade of honeymoon—the best years ever in the relations between the two countries. It is not simply that, since Tiananmen, China had been targeted. In a retrospective view, the U.S.-orchestrated political, economic, and military sanctions against China had triggered the process of unilateralism that was to culminate with the forthcoming Soviet collapse. Thus, it is already after Tiananmen that the United States had apparently given up China as a partner to the containment of the Soviet Union—when no one did or could predict its disintegration—and not after the Soviet collapse 2 years later, when China's partnership was no longer needed. Primarily concerned about economic growth and global stability, China was about to face the shaping of a new international order where only one voice ultimately counted—that of Washington.

It was China's conflicting interests in the Middle East, its internal challenges after the Tiananmen confrontation, and its deteriorating relations with the United States and the Western countries that had determined its hesitation and ambivalence with regard to the second Gulf conflict. Initially it appeared like a clear-cut case. While Beijing firmly condemned Iraq's aggression and insisted on its withdrawal from Kuwait, its instinctive response was that the conflict should be peacefully settled without the interference of external powers and "within the scope of the Arab countries."[15] This attitude reflected Beijing's time-honored Maoist and traditional policy that external powers, certainly the superpowers, had used and even fueled regional conflicts in order to consolidate their presence and promote their interests. Therefore these powers, and international organizations such as the UN that they allegedly control and manipulate, should by no means become involved in regional armed confrontations.

Soon, however, Beijing realized that the world—as well as China—had changed. The PRC itself was now a permanent member of the UN Security Council, charged with forging world peace. Ten years of successful economic reforms had injected a good deal of self-confidence into the PRC's internal and international behavior, winning a good deal of respectability and legitimacy both at home and abroad, at least until June 1989. Viewed in this perspective, the Iraqi invasion was a godsend that could be used to overturn the negative outcomes of Tiananmen. Managing the Gulf War correctly, the PRC could buy a ticket to the great powers' club by making its own contribution to the settlement of the conflict; by proving its indispensability as a UN Security Council veto power holder; and by forcing Washington and its allies to revoke, or at least relieve, their imposed sanctions on China. Nearly 20 years after its admission to the UN, this was surprisingly the first time that the Chinese could, and would, actually play a significant role in international crises, something they had consistently avoided in the past.[16] This, however, was easier said than done.

To begin with, Beijing had to make the crucial choice between its traditional *zuoshi* (sit and watch) noninvolvement policy and involvement. Given China's domestic and international predicaments, involvement was not only imperative in the negative sense (China could no longer escape its international obligations) but also and primarily in the positive sense (namely the opportunities that the conflict presented for China). Yet this was the first, and easy, step in Beijing's decisionmaking process. The next step was much more complicated and tricky. Beijing had to choose among three options. The first was to identify with the Iraqi side in the conflict. This, however, would have forced the Chinese to alienate some of the Arab countries, something they had been trying hard to avoid and that could have led to undermining their relations with some of their main Middle Eastern allies (such as Egypt, Iran, Saudi Arabia, and Kuwait—the victim itself). Furthermore, such a choice would have also contradicted Beijing's aspirations to reverse the Western verdicts—not to mention the premise that Iraq was doomed militarily. Beijing's second option was to support the U.S.-Western side in the conflict. This, however, would have created an uneasy association between China and former colonialism and would have tarnished China's self-determined image as *the only* representative of the Third World among the world's board of directors. Such a choice also could have undermined China's credibility in the Middle East. Finally, Beijing picked the third option, steering midway between the other two. Rather than voting for the resolution to intervene by force or against it, the Chinese, in a typically Confucian way, decided to abstain. Thus, China did become involved in the crisis but only in a limited sense, adopting the minimal commitments that would make all of the parties at least partly satisfied. China's decisionmaking with regard to the Gulf War reflected a clear distinction between words and deeds, multilateralism and unilateralism.

For example, Chinese media as well as classified speeches and documents treated Saddam Hussein, named "the little (or regional) hegemonist," much more leniently than President Bush, named "the big (or global) hegemonist." While Iraq, according to the Chinese, simply used the disintegrating international system and the weakening of the Soviet Union to push forward its territorial ambitions, the United States used the opportunity to consolidate its position as the predominant superpower. In other words, at the beginning Beijing tended to interpret the war as a particular indication of the gradual breakup of the bipolar international system and its replacement by a multipolar one.[17] It did not take long, however, for Beijing to realize the longer-term universalistic implications of the war. Given the collapse of East European (as well as Mongolian) socialism and the Soviet difficulties, bipolarity had indeed begun to crumble yet in favor of *unipolarity* rather than *multipolarity*. China's decisionmaking reflected no perceived illusions. European leaders were hardly consulted during the war.[18] All Chinese diplomatic efforts were channeled to Washington whose determination to use military force had been unshakable. As in a Greek tragedy where the end is known from the very beginning, there was nothing the Chinese could have done but to make the best out of their own predicaments.

This fundamental choice dictated China's decisionmaking process. It distinguished between two kinds of issues: those directly related to Iraq's aggression and those that involved external military intervention. Based on this distinction and despite repeated Iraqi requests, the PRC voted for all 11 UN Security Council relevant resolutions *including* Resolution 661 (that imposed economic and military sanctions against Iraq). Yet instead of supporting (or opposing) Resolution 678 (adopted on November 29, 1990, and authorizing the use of force to expel Iraq's forces from Kuwait), Beijing decided to abstain. There was no other way after China had already voted for several resolutions that imposed sanctions (661), a naval blockade (665), and an air blockade (670) against Iraq.[19] Since the Chinese could not vote for military intervention even under UN auspices—invoking bitter memories of the Korean War—least of all veto it, this abstention was enough to put Sino-American relations on track again as well as to avoid antagonizing China's Middle Eastern friends.[20] To be sure, Beijing's decision had been preceded by several meetings intended to guarantee its expected outcome.

A shrewd bridge player, Deng Xiaoping had launched a brinkmanship policy without any trump card, and Washington blinked first. In September PRC MFA Qian Qichen met U.S. Secretary of State James Baker at the UN for the first time after Iraq's invasion and again in Cairo in early November. Washington had been concerned that China would veto a military action against Iraq. In the words of Richard Solomon, then Assistant Secretary for East Asia and Pacific Affairs at the State Department:

> After the Iraqi invasion of Kuwait, it was evident that if we were going to have a UN coalition, or at least the UN sanction of some collective effort to deal with Saddam's aggression, we would have to work with the Chinese, given their veto position on the Security Council. The Chinese basically took a passive position. They were very anxious to avoid setting a precedent on the use of force, or seeming to cooperate with us too

closely. It was in that environment that the State Department reactivated its [post-Tiananmen] dealing with the Chinese, at least at the assistant secretary level.[21]

Yet, a careful study would have shown that Beijing had never used its veto power, and it was inconceivable that it would, given the circumstances. Furthermore, China must have warned Iraq that by no means would it veto the use of military force given Iraq's breach of international law.[22] It is inconceivable that Washington did not know about it. Yet the pretension that China could use its veto had paved the ground for the lifting of the economic sanctions that affected not only China's modernization, but also some U.S. commercial circles. In return for not using its veto (which, most likely, it would not have done anyway), Beijing requested and expected, as a first step, a meeting between Foreign Minister Qian Qichen and President Bush in the White House the day after the voting. Still, although the UN resolution passed, the Chinese abstention irritated U.S. policymakers to the point of canceling the agreed meeting. It was only under intensive PRC pressure and after it appeared that China might have been *asked* by the United States to avoid vetoing the proposal rather than support it, that Washington gave in. The next day Qian Qichen became the first senior Chinese official to obtain a meeting at the White House since June 1989.[23] Thus, even before military action was initiated to resolve the Gulf conflict, Beijing had won its most significant victory. Western Europe had outrun the United States in the race to lift the sanctions against China even before the voting took place. By the end of December, Japan had resumed its held-up loans and financial assistance to China. Finally, despite strong congressional opposition, Washington approved, yet again, China's Most-Favored-Nation trading status. As a side benefit, the Gulf War and China's role in it also helped to distract international public opinion from China's harsh treatment of "counterrevolutionaries" accused of fomenting the Tiananmen demonstrations.[24]

The second Gulf conflict had presented China with an opportunity to play an active and independent role in an international issue and restore its distinctiveness compared to the other powers. Yet, instead of underscoring moral values, the nonuse of military power to settle conflicts and its own indispensability as a permanent member of the UN Security Council, China once again retreated into the warm comfort of traditional inaction (or lack of effort, *wuwei*), hoping (as the term implies) that all would be done anyway. Beijing's abstention was, indeed, a demonstration of independence, but one that reflected passivity rather than activism. According to a Hong Kong journal, Deng Xiaoping (who undoubtedly had authorized this abstention and had definitely known about it) reportedly commented: "When I saw on the television that Qian Qichen unhurriedly raised his hand in 'abstention,' I nodded to him and saluted him. By holding up his hand, he again showed the whole world that China has a decisive say in solving major disputes in the world. Our foreign policy is firm and principled."[25]

Since this was supposedly said to his bridge partners and elder Chinese Communist Party (CCP) leaders, it reflects the extent to which China's leadership has been out of touch with reality or refused to face it. For by abstaining, Qian in fact indirectly supported a resolution that subjugated UN interests to those of the United States, not essentially different from the process that had led to the Korean War that Beijing has always condemned.

China failed to utter a single word for strengthening UN peacekeeping, let alone propose an alternative multilateral nonhegemonic peacekeeping force to deal with the Persian Gulf crisis. China's actual role can be better described as an unprincipled fishing expedition in troubled water, making the best of all possible worlds and seeking an escape route from international sanctions. China failed to capture the high moral ground in allowing the Security Council to legitimate an American war in the Persian Gulf.[26]

Although the Chinese had better access to all parties compared to other powers, they deliberately avoided any attempt to mediate,[27] something that could have won them international prestige, respect, and political capital—all lost since June 1989. One could speculate that, deep down, Beijing deliberately failed to prevent the war also because of the rare opportunity to observe and glean intelligence about Washington's military capabilities. Beijing's second Gulf War decisionmaking ultimately had been determined by two time-honored predispositions: tradition, and fear of the United States. In these respects, China won the short-term tactical and particularistic battle but lost the long-term strategic and universalistic war.

The outcome of the war reinforced Beijing's earlier suspicions that Washington had deliberately exacerbated the Gulf War and used it in order to increase its control over the Middle East, another step toward world domination. In explicit terms, remindful of similar concerns from the 1940s (Germans) to the 1970s (Americans and Russians), Beijing considered U.S. presence and predominance in the Middle East as a threat and a stepping-stone directed against China. Mao's siege mentality had resurfaced. To a considerable extent, this U.S. threat had literally been China's hand-made. By raising its hand in abstention, Beijing, in fact, had paved the ground for the American buildup in the Middle East—and all its consequences. It could have been avoided, but at an exorbitant cost that China could not afford. China's mission has been, from now on, to combine and reconcile its dialectical relationship with the United States as a rival and a partner at the same time. One way has been to upgrade its military capabilities.

It is conventionally accepted that the alleged phenomenal success of Operation DESERT STORM had "profoundly" shocked China's military (and political) leadership so as to trigger a Chinese-style Revolution in Military Affairs.[28] While the first Gulf conflict had been based on a low-tech protracted confrontation that fundamentally conformed to China's own outdated military doctrines and deficient hardware, the second Gulf conflict—based on high-tech operations and sophisticated equipment—allegedly betrayed at a stroke the miserable backwardness of China's defense system. I use the term "allegedly" for a number of reasons. One, as impressive as the U.S.-led offensive had appeared at the time, it later emerged that much of it had been public relations, and many supposedly advanced and sophisticated systems had failed to function properly. The Allies had, of course, enough firepower and technology to overcome Iraq (though never entirely) and thereby to impress the Chinese, yet their victory should be somewhat qualified, definitely in a retrospective view. Two, the Chinese did not need Operation DESERT STORM to reveal the "secret" of their own outdated military system.[29] There was no secret, and China's post-Mao leadership—not to mention all Western academics and observers without exception—had been well aware of China's military incompetence at least since its 1979 malfunction in Vietnam, if not before. It is likewise inconceivable that China's military leaders and intelligence services had been totally unaware of the outstanding sophistication of the Western and especially U.S. defense systems.[30] Finally, one has to take into account Beijing's tendency to inflate certain situations and threats out of all proportions. This "blow-up" has been intended either for internal consumption as an incentive to overcome domestic conservative and bureaucratic opposition or, more often, as a Sun Zi-style deception campaign intended for external consumption. Traditional Chinese dialectics stipulates that becoming powerful should begin with an awareness of being weak. Such awareness can be infused by constantly reiterating the adversary's power—which is what China has been doing in countless military publications since the early 1990s.

To be sure, there is no doubt that, objectively speaking, the PRC was weak militarily. Accorded the last priority among China's Four Modernizations, the so-called military "reform" undertaken by the PLA in the 1980s was marginal in the sense that it hardly touched on issues of technology and related military theory. By the time of the second Gulf War, the PLA had indeed undergone some reorganization, demobilization, and modernization of advanced weapons— yet on a limited scale. Its fundamental concept of traditional and conventional military competition with its adversaries had remained unchanged. The PLA still had a long way to go to achieve conventional parity, not to mention high-tech military technologies. In the meantime, Beijing used the Gulf War deliberately to underline and advertise its military weakness rather than vice versa.

Apparently, the impact of the second Gulf War on China's military was immediate. Within 1 year of the conflict, the Chinese had indeed "adopted" a revised national defense strategy—"winning a regional limited war under high-tech conditions" (*xiandai gaojishu tiaojianxia de juebu zhanzheng*). However, China often uses words as a substitute for action, and for a few years, it practically remained a slogan, a theory that has been displayed in numerous books and articles but never really internalized or implemented. A former Minister of Electronics, President Jiang Zemin, suddenly made a startling discovery. Summarizing the experience of the Gulf War, he stated that "military electronics has a bearing on national security" and "must be given first place." Indeed, since the early 1990s, China began to increase systematically and consistently its defense budgets and to buy large quantities of weapons and military technology, primarily from Russia, a source that had been unavailable for 30 years. Still, all these developments should be interpreted in terms of China's traditional strategy of relying on conventional and nonconventional low-tech weapons. Actual, to distinguish from rhetorical, adaptation to high-tech warfare has been far slower.

There is no way to transform any military system immediately, let alone China's conservative and bureaucratic colossus, in response to external (or internal) crises. In addition to problems in short-term defense allocation, China's military development processes are linked to long-term plans and concepts that can by no means be changed overnight. Therefore, it is only since the mid-1990s that Beijing sources have begun to underline the significance of asymmetrical warfare and to realize that in no way would China be able to withstand the United States in conventional terms. However, since the United States has become so totally dependent on computerized high-tech military systems—the ultimate lesson of the Gulf War was to use digital, space, and information warfare so as to disrupt the "brains" of U.S. military technology and make them ineffective. It is only since the late 1990s that the first actual results of this new Chinese military modernization drive have become visible, while its most important components, if any, probably remain invisible.

The visible aspects of China's defense modernization tell very little about China's actual defense modernization. Scores of books, articles, and Internet information about high-tech warfare mostly reflect foreign experience, primarily based on the 1991 Gulf War. They aim at making such warfare more familiar to the public, both civilian and military,[31] but by no means reflect China's actual defense modernization.

> The output is so vast that the proverbial unsuspecting visitor from Mars would be forgiven for thinking that the PLA is in the forefront of the dramatic changes taking place in how we think about and wage conventional war on Earth. It is not. Writing about and dissecting the RMA theoretically and conceptually, and actually being able to exploit it are two totally different endeavors. Whether the PLA succeeds in the second is another matter altogether.[32]

This vast literature has also been used to refute Washington's "China threat" allegations by implicitly or even explicitly underscoring the huge technological gap between China's obsolete military system and that of the United States. The establishment of scores of military academies, the dramatic increase in international military exchanges, and the "gradual but noticeable" increase in professional military education[33] are, at best, an initial step toward defense modernization and by no means a substitute for the real thing. It was only in early 1999 that the first major revision of the PLA operational doctrines since the mid-1980s was implemented, incorporating for the first time lessons of the Gulf War.[34] Some of these lessons have been adopted even later, and some not at all.

In sum, Beijing's reaction to the second Gulf conflict had indeed produced a more lenient U.S. policy toward China, but only for a short while. With the collapse of the Soviet Union, Washington has begun to perceive China as its main regional and even global rival. At the same time, the two countries' economies have become intertwined to the point of being mutually dependent. As for China's position in the Middle East, some observers believed that China's abstention that eventually facilitated the U.S.-led offensive in Iraq would have crippled its relations with the Arab countries, primarily with Iraq, and that its credibility would have suffered "serious damage."[35] This did not happen, nor could have, given the history of Sino-Arab relations. Earlier frictions have always been forgotten and forgiven (which is true of many countries). If Beijing did suffer a "serious damage," it was in economic and military relations. China failed to prevent Iraq's punishment, and the lucrative Iraqi arms market suddenly evaporated. In fact, China's arms exports in general have begun to decline since the early 1990s, not simply because the Iraqi arms market was blocked but mainly because of the poor performance of Chinese conventional weapons in the war. Overall market demand for Chinese weapons declined considerably afterward.

Yet the outcome of the war, in Beijing's perspective, has not been totally negative. Actually, China's smart policy of having the cake and eating it too had managed to differentiate itself from both the war coalition and the antiwar coalition. Despite their abstention, the Chinese continued to maintain good relations with Iraq and, in 1997, won a number of oil-sharing concessions that—once the UN-imposed sanctions were to be lifted—would contribute to reducing China's growing oil shortages. Some argue that China's abstention has damaged its image as a Third World leader. Yet, by the early 1990s China already had begun not only to dissociate itself from Third World developing countries, but also to successfully compete with them over Western capital resources. To be sure, notwithstanding their rhetoric, the Chinese have never been terribly fond of Saddam Hussein, nor of Islamic Iran, but they have been uncompromisingly fond of their interests. Thus, for example, they used the inter-Arab friction, Israel's restraint in the face of Iraqi SCUD missiles, and the war's propulsion of the Palestine problem to the top of the international agenda, to establish full diplomatic relations with Jerusalem on January 24, 1992—with virtually no Arab, Iranian, or Palestinian resistance.

For the Chinese, the second Gulf conflict was more complex and multifaceted than the first, since it involved an external armed intervention, though under UN auspices, which made it easier for them to go along with the coalition. This is why Beijing's decisionmaking concerning the second Gulf conflict could be termed "intermediate." In the next Gulf conflict, China had to manage an even more intricate and problematic situation.

THE THIRD GULF WAR: ADVANCED DECISIONMAKING

Unlike the previous two, the third Gulf conflict was essentially different. For one thing, while the first two had erupted suddenly, giving Beijing practically no time to prepare, the third conflict was brewing for months, if not years. For another, China's multilateral situation in 2003 was substantially different compared to 1990 and 1980. Tiananmen had been practically forgotten, if not forgiven, both at home and abroad. Reflecting growing self-confidence, China, which had by and large been cast aside after 1989, now has been recognized as an upcoming economic power and—at least from Washington's viewpoint—as a regional or even a global military threat. Finally, the global bipolar system that still had existed when the first two Gulf crises had taken place, has been replaced by a unipolar system whereby the United States plays the predominant global role, overruling and undermining international norms, procedures, and organizations. And, while in the past the Middle East had played a marginal role in China's actual, to distinguish from rhetorical, interests, by the early 2000s, Middle Eastern oil had become indispensable for China's continued economic growth. Arms sales, labor export, and construction services have declined. All these circumstances had to be taken into consideration by China's reaction to the third Gulf War.

One of Beijing's greatest concerns with regard to the forthcoming U.S.-led offensive against Iraq had been to ensure a steady oil supply, not only from Iraq but also from the entire Persian Gulf. Since the early 1980s and 1990s, when China's policymakers had failed to anticipate that China's energy needs would grow much faster than its (phenomenal) gross domestic product growth, China has become more and more dependent on Persian Gulf oil (see Table 2). A disruption of oil supply and/or a rise in oil prices could slow down China's impressive modernization drive and development.[36] It was only in the early 1990s, after the second Gulf War and at a rather late stage, that China began to diversify its oil resources. But China miscalculated and underestimated the amounts of oil it had to import, not to mention the belated realization that ultimately there is no substitute to Middle Eastern oil and that most alternative sources of supply are either unreliable in the long-run and of limited potential or politically risky. Admittedly, Middle Eastern oil is no exception. Persian Gulf oil-producing countries are usually perceived as unstable, violent, and subject to international power politics. Still, this is the only part of the world that contains plenty of easily accessible oil for all, China included. Indeed, in June 1997, Beijing committed $1.26 billion for the development of the al-Ahdab oilfield in Iraq, in an agreement for 22 years to become effective after the removal of the UN sanctions. At the same time, Beijing held negotiations aimed at signing additional concessions (production sharing agreements [PSA]) related to at least three other oilfields. China's 50 percent share in their combined output could, if accomplished, provide China with about one half of its annual oil imports in that year, an enormous amount. This is why the PRC was so interested in the lifting of the UN sanctions imposed on Iraq following the second Gulf War. Military interests have also been involved. Northern Industries Corporation (NORINCO), an armament industrial conglomerate, has been partner to the oil agreement in Iraq.[37] Unlike the two previous Gulf crises that had affected Chinese interests moderately—the imminent war against Iraq could lead to a dramatic deterioration in China's energy balance.

Exporter	1997	1998	1999	2000	2001
Iran	2,756,718	3,619,989	3,949,291	7,000,465	10,847,008
Iraq	239,010	607,352	974,155	3,183,182	372,056
Kuwait	68,790	282,285	330,443	433,428	1.459,823
Oman	9,033,023	5,793,430	5,020,825	15,660,840	8,140.355
Qatar	80,752	—	—	1,598,902	1,325,553
Saudi Arabia	499,908	1,807,618	2,496,968	5,730,211	8,778,376
UAE	48,438	514,506	—	430.474	649,766
Yemen	4,055,011	4,043,151	4,132,183	3,612,424	2,286,946
Sub-Total	16,781,650	16,668,331	16,903,865	37,649,926	33,859,883
Percent of Total	47.31%	61.01%	46.17%	53.58%	56.19%

Source: *Almanac of China's Foreign Economic Relations and Trade,* various years.

Table 2. China's Crude Oil Import from the Middle East, 1997-2001 (In Million Tons and Percent).

China began to increase its oil imports frantically, mainly from Africa and Russia, in February 2003, just 1 month before the outbreak of the war. This effort coincided with calls to raise China's strategic oil reserve to about 90 days' net import. As one of China's countermeasures to deal with the coming war between the United States and Iraq a National Energy Commission was established. The new organization was set up to design a national energy and oil security plan, map out an integrated security and development strategy, adjust the structures of energy production and consumption, and reduce reliance on crude oil and natural gas.[38] Articles warned that the anticipated war would ignite an oil panic that would not only lead to a rise in oil prices but would also affect trade, labor export, and construction services—essential Chinese activities in the Persian Gulf since the 1980s. "Once war begins between the United States and Iraq, it will not only be disadvantageous to China's manufacturing industry and export trade, but it will weaken the momentum of economic growth."[39] In fact, by July 2002, over 40 Chinese companies, with over 200 staff members, already had been engaged in power, locomotive, petroleum, and chemical industries in Iraq.[40] This, however, was a relatively minor problem.[41] More disturbing was the Chinese belief that, in addition to seizing Iraq's oil, Washington had broader geopolitical objectives. Beijing perceived the overthrowing of Saddam Hussein as the key for suppressing Iran, for restraining Syria, for promoting democracy, and fighting terrorism. "Regardless of how a new war in the Gulf is concluded, the United States inevitably will establish a more solid and powerful military base in the Middle East. This will have a disadvantageous impact on China's strategic development in the Asia-Pacific region and the strategic balance in the region will come under definite assault."[42] In fact, long before the war the United States had begun to shatter China's foothold in Iraq. Hua Wei Technology, a large telecom equipment manufacturer based in Shenzhen (and reportedly linked to the military), had been alleged by Washington to have installed fiber optic cables to help upgrade Iraq's air defenses. While denying the allegations, the company nevertheless pulled out of a US$28 million Iraqi mobile telephone project that had won UN approval.[43]

Given these interests, China's official best choice was, as in the past, to settle the conflict peacefully, by consultation, and preferably under UN auspices. If inevitable, war should be carried out and commanded by the UN, to be over as soon as possible. The Chinese obligations in Iraq had reached more than US$7 billion. "If the war ends quickly, the situation will also rapidly return to calm, which will bring economic and trade opportunities to China . . ."[44] Soon after the end of the second Gulf crisis, China began to consistently urge the removal of the UN-imposed sanctions against Iraq. The persistence of the sanctions damaged China's "economic and trade opportunities" not to mention military ones. Based, perhaps, on more intimate intelligence, Beijing has maintained all along that Iraq did not have any weapons of mass destruction. Foreign Minister Qian Qichen was occasionally quoted: "I can say that what was discovered has been destroyed. And there are doubts about the existence of those [arms of mass destruction] which have not been discovered yet."[45] But there was practically nothing China could, or would, have done.

Unlike the second Gulf conflict when China—as a permanent member of the UN Security Council—could have made a difference, Washington's decision to sidestep the UN in the third Gulf conflict had led to the marginalization of Beijing (and a few others).[46] This time the Chinese could not threaten Washington even implicitly, let alone explicitly, and could in no way prevent the U.S.-led offensive. Its veto power had become useless. By the beginning of the 21st century, the PRC, notwithstanding its phenomenal economic achievements (or partly because of them), had been incapacitated politically. Though the similarities are limited, one could say that its obsession with economic growth and increased interaction with the world economy in terms of both input and output, have led China to a gradual "Japanization." By putting the economy at the top of its agenda since the early 1980s, Beijing has, by necessity, inevitably restricted its ability to assert itself politically and militarily. Yet its clipped wings by no means implied that China could or should have remained indifferent to the evolving war. Too much had been at stake, and Beijing was compelled once again to carefully make the right policy choice.

From the very beginning, Beijing had kept a low profile with regard to the conflict and especially to Washington's plans. Although the Chinese intrinsically opposed the war in general, and the United States in particular, their attitude was more restrained than France's, Germany's, and Russia's—that articulated the same opposition yet in a vocal, official, and consistent way. Although China backed a French-German-Russian proposal that would extend weapon inspections in Iraq and thereby prevent, or at least delay, the war, this was no more than a tactical move. Beijing had had no intention of counteracting Washington's determination to attack Iraq, considered practically as a *fait accompli*.[47] No endeavor was made to mobilize the masses against the United States or to provoke anti-American feelings among the crowds, as Beijing had done following the bombing of its embassy in Belgrade. Unlike the mass demonstrations against the war held throughout the world, only a few took place in China and with a relatively small number of demonstrators.[48] China deliberately excluded itself from the "antiwar axis," perhaps on the basis of earlier understandings with Washington.[49]

Some of the reasons for Beijing's behavior are obvious, some are not. Obviously, and based on their reading of the second Gulf War, China had been disillusioned as to the successful outcome of the U.S.-led offensive. This time there was no reason for the leadership to be "profoundly shocked." Although there is no concrete evidence, and will never be, it appears that Beijing was interested in a U.S.-led military intervention on the assumption that in this way the crisis would be brief and least disruptive. Eager to sustain regional and international stability for the sake of its economic growth and now substantially dependent on Persian Gulf oil, China could not afford the time nor the patience for a protracted struggle. If crises could not be solved peacefully, then an immediate and swift military action was preferable, mainly from an economic point of view. Yet, from a strategic and political perspective, as long as the United States is occupied in Iraq, China and North Korea are off the hook.[50] But Beijing's implicit interest in a military action could imply another, even less obvious reason. Based on the lessons of the second Gulf crisis, Beijing must have expected another free, not less and probably much more educational, display of advanced U.S. military power, far away from PRC territory. If, as conventional wisdom says, the American military performance in the second Gulf crisis had become a, or rather the, major booster for the Chinese defense modernization, the anticipated American performance in 2003 promised to be an even better gift. Furthermore, too much emphasis on the need to settle conflicts peacefully could have undermined Beijing's refusal to rule out the use of military force to solve the Taiwan problem. Time and again Washington has provided China with precedents that some problems could be solved only by force, thereby legitimizing a Chinese use of force against Taiwan in the future.[51] Outwardly concerned about the war, Chinese officials, both civilian and military, could by no means betray any interest in the war either publicly or privately.

This is why Beijing has so easily, almost eagerly, given up some of its fundamental principles. At best, post-Mao China has always insisted on a peaceful settlement of conflicts, preferably by the parties concerned and without any external intervention. At worst, Beijing has been ready to accept exogenous collective action, but only under UN auspices. In these respects, the first Gulf conflict had been the simplest for China to deal with, since there had been no external involvement and China could sit, watch, and make money. The second Gulf conflict was more complex because of the external involvement, but the Chinese still excused themselves, for the military action had been sponsored by the UN. In these terms, the third Gulf conflict has been the worst, from China's standpoint, as it involved external military action, not under UN but under U.S. auspices, usually an unacceptable proposition. Still, Beijing had to weigh its interests against the odds and finally come up with a decision. This is why China's response to the third Gulf conflict should be called "advanced" decisionmaking.

Despite its friction with Washington and its refusal to deploy troops in Iraq, it seems that China's lower-key attitude toward the offensive, compared to the vocal Russian, French, and German opposition, has begun to pay off. Beijing pledged $25 million in humanitarian aid to Iraq, and indicated its readiness to write-off a big chunk of Iraq's debt, estimated at $5.8 billion. In return, the Chinese expect to take an active part in Iraq's reconstruction after the end of the war, and to activate their vital oil production sharing agreements. As a result of its cautious policy, the PRC is not on the U.S. list of countries such as Russia, France, and Germany that are excluded from bidding for the reconstruction of Iraq. Although its agreements have been suspended, by March 2004 two Chinese companies have secured deals, both in the telecommunications sector. A month earlier, a small Chinese team started preparations for reopening the PRC embassy in Baghdad, seriously damaged during the war.[52] It is also quite possible that Washington's readiness to overlook Beijing's opposition to the Iraqi War also reflects its recognition of China's crucial role in dealing with the North Korean crisis.[53]

While it might be too early to estimate the military lessons that Beijing has drawn from the third Gulf War, initial conclusions can still be reached.[54] Perhaps the most important is that, contrary to the lessons of the 1991 and 1999 offensives in Iraq and Kosovo, air power and long-range precision strikes alone are not sufficient to prevail in an armed conflict. Ground forces are still essential to overcome an enemy, yet not in the traditional Chinese sense. An additional lesson drawn by the Chinese includes the integration of psychological warfare with air and rapid ground operations directed at the enemy leadership, its ability to communicate and willingness to fight.[55] The war also has reinforced PLA *plans* to improve weapons mobility and firepower and accelerate acquisition of information technology and advanced command, control, communications, computer, intelligence, survelliance, and reconnaissance (C[4]ISR) systems to upgrade joint operations capabilities and interservice cooperation and integration. Yet reading PLA reports and commentaries on modern warfare, one can get the impression that almost nothing has been *done* so far, and that military reform has yet to be launched by using the "historic opportunity" or "strategic opportunity." Commenting on the lessons of the war in Iraq, Dr. Zheng Yanping of the Military Science Academy underscored in bold terms the main reasons for China's failure to modernize its military system:

> It is fair to say [in July 2003, a quarter of a century after the beginning of China's so-called defense reform!] that our military is still a surface combat military. Our strategic thinking is inevitably imprinted with much traditional thinking. The backwardness in thinking is fundamental backwardness. To push forward with Chinese characteristics, we should be brave in emancipating ourselves from those modes of thinking that once brought us glory, but have proved outdated by practice. . . . We should bring about new military theories with new thinking so as to bring about and guide military reform with Chinese characteristics.[56]

CONCLUSION

While all three conflicts discussed in this chapter similarly involved Iraq and Saddam Hussein's intransigence against Iran, Kuwait and the United States (or UN), they led to different international responses. The first conflict, the Iran-Iraq War, had failed to draw world attention until 1987, when the vital traffic in the Hormuz Straits was threatened. Only then did the powers intervene to cut short the confrontation. The second war, that had begun with Saddam's invasion of Kuwait, immediately triggered a Western response in the form of a military offensive led by the United States yet under UN auspices. International response to the third war and Saddam's refusal to allow an in-depth inspection of Iraq's (actual or virtual) nonconventional weapons had provoked a unilateral

U.S. response that bypassed not only the UN but also some of its European allies, as well as Russia and China.

Despite these variations, China's decisions on all three conflicts have been basically identical, although one could detect different attitudes by diverse groups representing different interests (including the military, the economy, the Foreign Ministry, and the CCP). Fundamentally and officially, post-Mao China always has been interested in maintaining regional and international stability as a precondition for its modernization and economic growth. Consequently, Beijing has consistently upheld the principle of settling conflicts peacefully through consultation between the parties concerned and without external interference. Yet, while paying lip service to this principle, the PRC has become involved in the Iraqi wars and—directly (in the first) or indirectly (in the two others)—contributed to prolonging and extending their military dimensions. In all three, the Chinese have hardly lifted a finger to enable a peaceful settlement and have ultimately, though implicitly, gone along with the U.S.-led coalition's offensive, either within the UN framework or without. Given the growing U.S. hostility against China from the early 1990s and Washington's reiterated "China threat" theory, why have the Chinese opted to follow, unofficially of course, the U.S. rules of the game?

One conventional answer is that the Chinese have had no choice. Following the fall of the Union of Soviet Socialist Republics, Beijing's dreams of a multipolar international system were shattered as U.S. unilateralism began to gather momentum.[57] The Chinese could not afford to defy the United States primarily because of their economic dependence on the American market, but also because of their military inferiority. Apparently, this inferiority was irrelevant in terms of the three Iraqi wars, but as we shall see in a minute, it has been crucial for understanding Beijing's response. As Beijing has been trying to restore its image as a responsible and reliable partner to the United States following Tiananmen; its admission to the World Trade Organization; and its attempts to curb nonconventional arms proliferation, it could not but join—implicitly rather than explicitly—the U.S.-led coalition. All these explanations proceed from the basic assumption that the Chinese wanted very much to prevent the

Iraqi wars, but have been forced to give up active opposition to the offensive under the existing circumstances and constraints. Is that so?

A nonconventional explanation would suggest that, while these confrontations have entailed some short run nonessential losses for Beijing, their benefits have been more essential, far greater, and long run—less in economic terms, more in political terms, and mostly in military terms. This is evident in three wars. As we have seen, China's economic relations, trade turnover, labor exports, and construction services increased dramatically during the first Iraqi war. The second Iraqi conflict and the UN-imposed sanctions brought Sino-Iraqi economic relations almost to a standstill, though, in view of the overall size of China's foreign economic relations, the damage has not been so bad. The same goes for the third conflict. Some Chinese companies already have been allowed by Washington to take part in Iraq's post-war reconstruction, including nearly US$3 billion agreements on the construction of some 20 electric power stations, signed before the war.[58] Still, China's 1997 oil production agreements with Baghdad remain suspended, and the prospects of Iraq becoming a major oil supplier to the PRC still look poor. Yet alternative large-scale suppliers such as Saudi Arabia, Iran, Oman, and others have already stepped in. In economic terms, the damage has been limited; in some sectors, marginal.

In political terms, China's management of the Iraqi crises has produced much better outcomes. Despite its cooperation with both belligerents in the Iran-Iraq War, China managed to maintain its credibility and good relations with the other Middle Eastern countries. Furthermore, at the beginning of the conflict, no one had taken the Chinese seriously as noteworthy players in international affairs. Yet by the end of the war, and from then on, Beijing has had to be taken into consideration as an actual (as well as potential) conventional (and nonconventional) arms supplier. Its arms proliferation policies have become an issue—and a bargaining card—in its relations with the United States. Covered by an implicit alliance with Washington, Beijing has used the Iran-Iraq War to become actively and substantially involved in a region far away from its borders for the first time in its history. In a retrospective view, this should be considered China's first step on the long march to a world power status.

Along this march, China was crippled by Tiananmen. Iraq's invasion of Kuwait had provided the Chinese with a golden opportunity to recuperate. Beijing's agreement not to use its (virtual) veto power in the UN Security Council opened the gate for the offensive against Iraq and also led to the removal of the civilian sanctions imposed on China following Tiananmen. China used the conflict smartly to regain its international standing and recover from its isolation. However, these impressive achievements were short-lived. With the collapse of the Soviet Union, China has been designated by the United States as its main adversary, leading to a growing pressure on China and to a number of incidents including the bombing of the PRC Embassy in Belgrade, the interception of the EP3, and China's military exercises in the Taiwan Straits. All these incidents show in one way or another the military imbalance between China and the United States. It is in this field that China has ironically managed to gain the most from the three Gulf conflicts.

As we have seen, the first Gulf War was used by China to supply large amounts of weapons to both sides, thereby accumulating an additional income of several billions of U.S. dollars that were partly, or mostly, channeled to feed China's defense reforms. Moreover, although little has been published about it, the Iran-Iraq War provided an unprecedented large-scale testing ground for Chinese-made weapons, substantially more extensive both in scope and in time than the 1979 experience in Vietnam. It is quite possible that the two consecutive confrontations produced the ultimate evidence of the poor performance of China's military hardware. It is no accident that Chinese arms transfers have declined considerably following the war. Even more important is the role of the Iran-Iraq War in underlining the urgent need to modernize the Chinese defense system drastically. How to do it? In which direction to go? It is Washington that offered the answer in the second and third Gulf conflicts.[59]

It is inconceivable that Beijing was unaware of the advanced American military technology that had been sporadically displayed well before the second Gulf conflict, but this awareness was shallow at worst and theoretical at best. The amazing military performance of the United States in Iraq has twice exposed its state-of-the-art technological and conceptual practical edge, thereby producing a model for the Chinese, either to follow or to prepare against. A good deal has been written on the impact of the second Gulf War on the Chinese defense modernization and military thinking. Although we do not know how far this modernization has actually gone, we do know confidently that it is the U.S. performance in the second Gulf War that had triggered China's Revolution in Military Affairs. If Beijing had been interested implicitly in enabling the second Gulf War primarily for political reasons, but perhaps also (if we attribute any sophistication to China's decisionmakers) for military ones, all the more so with regard to the third Gulf War. China's low-profiled "opposition" to the war in fact suggests that Beijing has been almost eager (again implicitly) to watch a repeat performance and to draw its own military lessons.

Viewed in a wider perspective of a long-run Sino-American rivalry, rather than in a narrower Iraqi-American perspective, the U.S. exposure of its military sophistication in 1991 and 2003 has been a duplicated mistake. By employing such advanced military power against such a primitive enemy (that could have been dealt with in other, more traditional ways), Washington, in fact, unnecessarily and impulsively betrayed military technologies, systems, and methods that could have otherwise surprised an adversary like China in case of a violent confrontation between the two. Now, China could learn how to cope with the U.S. military strength and weaknesses, while the United States will have to upgrade its military system at a huge cost in order to retain its edge. Strategically, the Chinese are undoubtedly also pleased to see the United States stuck in Iraq and forced to reduce its military presence in East Asia.[60]

From Beijing's perspective, all these conflicts have signified a change, occasionally sudden, in the regional and international system that affects China's interests, at least indirectly. Beijing's response to these conflicts has been determined by a number of inputs including the time dimension, the degree and location of the conflict, and its impact on its interests both in the negative sense (threats) and in the positive sense (opportunities). In these respects, Beijing's decisions have been made under stress of time and threatening implications, yet in a region far away and in a situation that has also offered opportunities. Ultimately and even under stress, China's decisionmaking process has been rational, representing a sophisticated balancing act of pros and cons.

Had similar conflicts taken place in a nearby region or along its borders, the Chinese reaction could have likely been substantially different. In such a case, Beijing would have done its utmost—possibly as far as using its veto power—to actively contain the conflict as soon as possible. This was the case with the bombing of the PRC embassy in Belgrade, the EP-3 incident, and it is the case with North Korea. In the case of Iraq, on the other hand, the extension of the conflicts has paid Beijing handsome dividends—primarily in political, military, and even economic terms.

ENDNOTES - CHAPTER 7

1. I am grateful to Mr. Itamar Livni for his research assistance.

2. See, for example, Lu Ning, *The Dynamics of Foreign-Policy Decisionmaking in China*, 2nd ed., Boulder, CO: Westview, 2000; David M. Lampton, ed., *The Making of Chinese Foreign and Security Policy in the Era of Reform*, Stanford: Stanford University Press, 2001; and Carol Lee Hamrin and Suisheng Zhao, eds., *Decisionmaking in Deng's China: Perspectives from Insiders*, Armonk, NY: M. E. Sharpe, 1995.

3. The following is based on Yitzhak Shichor, *The Middle East in China's Foreign Policy 1949-1977*, Cambridge: Cambridge University Press, 1979. See also Mohamed Bin Huwaidin, *China's Relations with Arabia and the Gulf 1949-1999*, London: RoutledgeCurzon, 2002; Lillian Craig Harris, *China Considers the Middle East*, London: Tauris, 1993.

4. *Beijing Review*, No. 39, September 29, 1980, p. 5.

5. Yitzhak Shichor, "China and the Role of the United Nations in the Middle East: Revised Policy," *Asian Survey*, Vol. XXXI, No. 3, March 1991, pp. 255-269.

6. Data from Richard F. Grimmett, *Conventional Arms Transfers to the Third World, 1982-1989*, CRS Report to Congress, Washington, DC: Congressional Research Service, June 19, 1990. See also R. Bates Gill, *Chinese Arms Transfers: Purposes, Patterns, and Prospects in the New World Order*, Westport, CT: Praeger, 1992, pp. 93-113.

7. Yitzhak Shichor, "The Year of the Silkworms: China's Arms Transactions, 1987," in Richard H. Yang, ed., *SCPS Yearbook on PLA Affairs 1987*, Kaohsiung: Sun Yat-sen Center for Policy Studies, National Sun Yat-sen University, 1988, pp. 155-156.

8. Lu Ning, pp. 122, 137, 154-155, n. 13, 14, 158, n. 62.

9. AFP, Hong Kong, August 5, 1987, in *Foreign Broadcast Information Service* (FBIS)-CHI, August 5, 1987, p. A2; *Washington Post*, September 27, 1987.

10 . Much of the information in this section is drawn from Yitzhak Shichor, "China and the Gulf Crisis: Escape from Predicaments," *Problems of Communism*,

Vol. XL, No. 6, November-December 1991, pp. 80-90. See also J. Mohan Malik, "Peking's Response to the Gulf Crisis," *Issues and Studies*, Vol. 27, No. 9, September 1999, pp. 107-128; Lillian Craig Harris, "The Gulf Crisis and China's Middle East Dilemma," *The Pacific Review*, Vol. 4, No. 2, 1991, pp. 116-125.

11 . On China's relations with Kuwait, see also Hashim S. H. Behbehani, *China's Foreign Policy in the Arab World 1955-1975: Three Case Studies*, London: Kegan Paul International, 1981, pp. 189-232.

12. Data from China's *Almanac of Foreign Economic Relations and Trade, 1992-1994*, pp. 790-804. More details in Yitzhak Shichor, "China's Economic Relations with the Middle East: New Dimension," *China Report*, Vol. 34, Nos. 3-4, 1998, pp. 419-439. About 5,000 Chinese workers were evacuated successfully from Kuwait after Iraq's invasion.

13. Data from Grimmett, *Conventional Arms Transfers to the Third World, 1983-1990*, pp. 64, 67; and Daniel L. Byman and Roger Cliff, *China's Arms Sales: Motivations and Implications*, Santa Monica: RAND, 1999, p. 51.

14. R. Bates Gill, *Chinese Arms Transfers*, pp. 105-113.

15. Gu Yuqing, "Qian Qichen Says That China Is Gravely Concerned over Iraq's Invasion of Kuwait," *Renmin Ribao (People's Daily)*, hereafter *RMRB*, August 5, 1990, in FBIS-CHI, August 6, 1990, pp. 14-15.

16. Shichor, "China and the Role of the United Nations."

17. "Li [Peng]: Gulf Crisis Should be Resolved by Peaceful Means," *Beijing Review*, September 10-16, 1991, p. 4.

18. Michael B. Yahuda, "China and Europe: The Significance of a Secondary Relationship," in Thomas W. Robinson and David Shambaugh, eds., *Chinese Foreign Policy: Theory and Practice*, Oxford: Clarendon Press, 1994, p. 282.

19. Richard Clutterbuck, *International Crisis and Conflict*, New York: St. Martin's Press, 1993, pp. 186-187.

20. According to a senior UN Western diplomat, the PRC representative was always present in the discussions of the five UNSC permanent members who drafted the resolutions regarding the Gulf crisis, but he "was often silent." Craig Harris, *China Considers the Middle East*, p. 318, n. 4. A Latin American diplomat characterized China's behavior in the Security Council: "They never take part in the give and take of preparing resolutions... If they can, they let others to weave together a resolution, then say they can live with it. They do not waste any time on things that are not fundamental to their interests." Quoted in David M. Lampton, *Same Bed, Different Dreams: Managing U.S.-China Relations, 1989-2000*, Berkeley: University of California Press, 2001, p. 168. China was the only Security Council member to abstain. Twelve voted for the resolution and two—Cuba and Yemen—against.

21. Nancy Bernkopf Tucker, ed., *China Confidential: American Diplomats and Sino-American Relations 1945-1996*, New York: Columbia University Press, 2001, p. 453.

22 . Craig Harris, *China Considers the Middle East*, p. 250.

23. This section is based on James Mann, *About Face: A History of America's Curious Relationship with China, from Nixon to Clinton*, New York: Vintage Books, 2000, pp. 248-250.

24. Tai Ming Cheung, "Silence in Court: Gulf War Mutes Foreign Concern over Dissident Trials," *Far Eastern Economic Review*, January 31, 1991, p. 14.

25. Quoted by Michael Yahuda, "Deng Xiaoping: The Statesman," in David Shambaugh, ed., *Deng Xiaoping: Portrait of a Chinese Statesman*, Oxford: Clarendon Press, 1995, pp. 157-158.

26. Samuel S. Kim, "China's International Organizational Behaviour," in Robinson and Shambaugh, eds., *Chinese Foreign Policy*, p. 423, 424.

27. Qian Qichen's visit to the Middle East in November 1990 was described retroactively as an effort to mediate, but earlier the trip had been called "a fact-finding mission." If indeed the visit represented a Chinese effort to mediate—an unlikely possibility—nothing happened. "Qian Drums Up Peace in Gulf," *Beijing Review*, November 19-25, 1990, p. 4.

28. David Shambaugh, *Modernizing China's Military: Progress, Problems, and Prospects*, Berkeley: University of California Press, 2002, pp. 1-3; Nan Li, "The PLA's Evolving Warfighting Doctrine, Strategy and Tactics, 1985-1995: A Chinese Perspective," *The China Quarterly*, No. 146, June 1996, pp. 456-458; You Ji, *The Armed Forces of China*, St. Leonard: Allen & Unwin, 1999, pp. 11-15. This claim is repeated over and over again.

29. One of the first—and few—to point it out was Ellis Joffe, "China After the Gulf War," SCPS Papers No. 7, Kaohsiung: Sun Yat-sen Center for Policy Studies, National Sun Yat-sen University, May 1991, pp. 8-9. See also Ellis Joffe, "The Chinese Army: Coping with the Consequences of Tiananmen," in William A. Joseph, ed., *China Briefing, 1991*, Boulder, CO: Westview, 1992, pp. 54-55.

30. On China's efforts to collect information—and technologies—in the United States, see Huo Zhongwen and Wang Zongxiao, *Guofang Keji Qibaoyuan ji Huoqu Jishu* (*Sources and Techniques of Obtaining National Defense Science and Technology Intelligence*), Beijing: Kexue Jishi Wenxuan Chubanshe, 1991. English translation in *http://www.fas.org/irp/world/china/docs/sources.html*. See also Nicholas Eftimiades, *Chinese Intelligence Operations*, Annapolis: Naval Institute Press, 1994; and Kenneth deGraffenreid, ed., *The Cox Report: U.S. National Security and Military/Commercial Concerns with the People's Republic of China*, Washington, DC: Regenery Publications, 1999.

31. Dennis Blasko, "PLA Force Structure: A 20-Year Retrospective," in James C. Mulvenon and Andrew N. D. Yang, eds., *Seeking Truth from Facts: A Retrospective on Chinese Military Studies in the Post-Mao Era*, Santa Monica: RAND, 2001, p. 81.

32. Ahmed S. Hashim, "The Revolution in Military Affairs Outside the West," *Journal of International Affairs*, Vol. 51, No. 2, Spring 1998, p. 440.

33. Thomas J. Bickford, "Professional Military Education in the Chinese People's Liberation Army: A Preliminary Assessment of Problems and Prospects," in James C. Mulvenon and Andrew N. D. Yang, eds., *A Poverty of Riches: New Challenges and Opportunities in PLA Research*, Santa Monica, CA: RAND, 2003, pp. 11, 15, 16-17.

34. Lonnie Henley, "PLA Logistics and Doctrine Reform, 1999-2009," in Colonel Susan M. Puska, ed., *People's Liberation Army after Next*, Carlisle: Strategic Studies Institute, August 2000, pp. 68-69. The definitive Chinese history of the war was not published until long after it ended. Academy of Military Sciences, Military History Research Department, *Haiwan Zhanzheng Quanshi* (*The Complete History of the Gulf War*), Beijing: Jiefangjun Chubanshe, 2001.

35. Craig Harris, *China Considers the Middle East*, pp. 251-252; and "Myth and Reality in China's Relations with the Middle East," in Robinson and Shambaugh, eds., *Chinese Foreign Policy*, pp. 339, 343.

36. Ding Gang, "Will Oil Crisis Erupt Again?" *Renmin Ribao*, November 25, 2002, in FBIS-CHI-2002-1125.

37. U.S. Energy Information Administration, "Country Analysis Brief: Iraq," February 1998, p. 4; "Middle East Oil and China," *Financial Times Middle East Energy*, No. 9, January 22, 1998, p. 4.

38. "Mainland Stocks Up on Strategic Oil Reserves," *Sing Tao Jih Pao*, Hong Kong, February 17, 2003, in FBIS-CHI-2003-0217.

39. Chen Mao-sheng, "U.S.-Iraq War Brings Economic Challenges for China," *Ta Kung Pao*, Hong Kong, February 27, 2003, in FBIS-CHI-2003-0227. See also Yang Jijian, Economic and Commercial Attaché in the PRC Consulate General in San Francisco, "War to 'Topple Saddam' Will Hit China's Economy," *Wen Wei Po*, Hong Kong, January 31, 2003, in FBIS-CHI-2003-0203.

40. *Xinhua*, February 13, 2003, in FBIS-CHI-2003-0213. A year later, there were only eight Chinese companies with branch offices in Iraq, with 23 Chinese employees. *Xinhua*, April 12, 2004, in FBIS-CHI-2003-0412.

41. As early as September 2002, in anticipation of the Iraqi crisis, Beijing had advised its personnel to leave Iraq. Urged to do so in early 2003 as tension in Iraq was growing, most economic and diplomatic staff had left Iraq by February 10, according to Lai Qinwei, the Commercial Attaché at the PRC Embassy in Baghdad. Zhang Lanhua and Liang Youchang, "Chinese Companies Evacuate Personnel from Iraq," *Xinhua*, February 13, 2003, in FBIS-CHI-2003-0213.

42. Lung Hua, "A New War in the Gulf and China's Strategic Security," *Hsin Pao*, Hong Kong Economic Journal, February 19, 2003, in FBIS-CHI-2003-0219; Gong Zhenxi, "The United States Adjusts Its Middle Eastern Strategy," *Liaowang*, No. 8, February 24, 2003, pp. 60-61, in FBIS-CHI-2003-0307; Zhang Shuguang, "What Is the United States Expecting from the Iraqi War?" *Renmin Ribao*, March 27, 2003, p. 13, in FBIS-CHI-2003-0327.

43. AFP, Hong Kong, September 5, 2001, in FBIS-NES-2001-0905.

44. Chen Maohong, "U.S.-Iraq War."

45. AP, December 26, 1997; *Reuters*, January 6, 1998.

46. "Beijing was likely relieved that after several weeks of haggling with the other UNSC members, the United States decided not to ask the UN Security Council to vote on a new draft resolution on the Iraq issue. Had a vote been taken, China probably would have abstained rather than veto to avoid a negative backlash in its relations with the United States." Bonnie Glaser, "China and the U.S. Disagree, but with Smiles," *Comparative Connections*, Vol. 5, No. 1, First Quarter 2003, p. 31.

47. On the power balance, see Xie Pu, Han Qinggui, and Li Zhishun, Research Fellows at the PLA Military Science Academy, "A Glance at U.S. and Iraqi Military Strength," *Renmin Ribao*, February 10, 2003, p. 3, in FBIS-CHI-2003-0210. See also Lin Hsing-chih, "It Would be Better To Follow the United States If the Tide Cannot Be Turned," *Hsin Pao* (Hong Kong Economic Journal), February 6, 2003, in FBIS-CHI-2003-0206.

48. See also John J. Tkacik, Jr., and Dr. Nile Gardiner, "China Will 'Stand Aside' on Iraq," *Policy Research & Analysis*, Heritage Foundation, February 12, 2003.

49. "Bush's China Visit Is To Pave Way for Attacking Iraq," *Sing Tao Jih Pao*, Hong Kong, February 16, 2002, in FBIS-CHI-2002-0216. Some argue that Washington's listing of the Eastern Turkestan Islamic Movement as a terrorist organization (without much evidential basis) in August 2002 was extorted by Beijing in return for its promise not to use its veto if a vote in the UN Security Council would take place. Jabin T. Jacob, "China's Position on Iraq vis-à-vis UNSCR 1441," *China Report*, Vol. 39, No. 3, 2003, pp. 407-410.

50. Sources in Beijing believed that the Chinese absolutely did not want the U.S.-Iraq war to end swiftly, despite the economic costs to China. On Beijing's dilemma, see Lung Hua, "China's Conflicted Feelings Toward the U.S.-Iraq War," *Hsin Pao*, Hong Kong Economic Journal, March 26, 2003, p. A23, in FBIS-CHI-2003-0326.

51. Jasper Becker, "War Poses a Double Dilemma for China," *International Herald Tribune*, March 28, 2003.

52. AFP, "China Tempered Criticism on Iraq and Now Reaping Benefits," March 11, 2004. See also Chen Hong, "Who Will Foot the 'Bills' for Iraq's Reconstruction?" *Renmin Ribao*, Guangzhou South China News Supplement, April 16, 2003, p. 2, in FBIS-CHI-2003-0416.

53. Ralph Cossa, "Diplomacy Fails with Iraq; Is North Korea Next?" *Comparative Connections*, Vol. 5, No. 1, First Quarter 2003, p. 4.

54 . The following is based on FY04 Report to Congress, *Annual Report on the Military Power of the People's Republic of China*, May 2004, in *www.defenselink.mil*, accessed on August 8, 2004. See also Willy Lam, "Beijing Studies the U.S. War in Iraq," The Jamestown Foundation, *China Brief*, April 8, 2003; and Garret Albert,

Michael Chase, Kevin Pollpeter, and Eric Valko, "China's Preliminary Assessment of Operation Iraqi Freedom," *Chinese Military Update*, Vol. 1, No. 2, July 2003, pp. 1-4.

55. See, for example, Li Xuanliang, "The Iraq War Opens the Door to the Use of Information Technology for Psychological Warfare," *Xinhua*, May 13, 2003, in FBIS-CHI-2003-0513. Though the PLA has had military psychology courses in its academies for well over a decade, it was only in 1999 that an office of psychological warfare was established and only 2 years later, in 2001, that the PLA's first psychological warfare experts graduated from the Xi'an Political Institute. See Bickford, p. 12.

56 . Zhang Guoyu, "Seize Development Opportunities and Push Forward Reform Practices—Notes on 'How to Push Forward Military Reform with Chinese Characteristics' Special Discussion session," *Jiefangjun Bao*, July 22, 2003, p. 6, in FBIS-CHI-2003-0722.

57 . Although some Chinese still would argue differently, see Sun Zifa, "Unipolar World Does Not Work. Multipolarism Order of the Day," *Zhongguo Xinwen She*, March 20, 2003, in FBIS-CHI-2003-0320.

58 . Weng Yang, "Beijing Expert Says War in Iraq Will Have a Limited Impact on China's Foreign Trade and Economic Relations," *Zhongguo Xinwen She*, March 20, 2003, in FBIS-CHI-2003-0320; Economic Research Office under the Financial and Economic Committee of the National People's Congress, "How Much Impact the Iraq War Had on China's Economy," *Renmin Ribao*, August 4, 2003, in FBIS-CHI-2003-0804.

59 . See also Paul H. B. Godwin, "The PLA's Great Leap into the 21st Century: Implications for the U.S.," *China Brief*, Vol. 4, Issue 9, The Jamestown Foundation: April 29, 2004.

60. Willy Lam, "China's Reaction to America's Iraq Imbroglio," The Jamestown Foundation, *China Brief*, Vol. 4, No. 8, April 15, 2004, pp. 1-2.

CHAPTER 8

"DECISIONMAKING UNDER STRESS" OR "CRISIS MANAGEMENT"?: IN LIEU OF A CONCLUSION

Frank Miller and Andrew Scobell

Since its founding, the People's Republic of China (PRC) has had its share of crises; some of its own doing, some forced upon it by nature or external forces. At times, Chinese leaders have acted quickly to take control of a situation and resolve the issue at hand. At other times, Beijing has seemed incapable of even recognizing a crisis existed, much less indicating it knew how to respond. This dichotomy exists both for domestic and international crises, in times of strong unified leadership, and in times of divided leadership. The overarching question that the contributors to this volume have wrestled with is whether they could discern a pattern for how China handles crises. What can one learn from analyzing case studies of Chinese crisis management? Of particular interest to the contributors was the role of the People's Liberation Army (PLA) in managing crises. In each case study, to what extent was China's military involved? Was the PLA central, pivotal, peripheral, or irrelevant?

CRISIS? WHAT CRISIS!?

As the editors of this volume note in their introductory chapter, there was widespread consensus that China's leaders did not consider all of the cases in this volume as crises. Nearly all participants agreed that in most instances the crisis was short-term and often position-dependent. In other words, notification of an unexpected event, such as explosions at the Chinese embassy in Belgrade or the emergency landing of a foreign military aircraft on Chinese territory, may be viewed as a crisis at least for a particular desk officer at the Foreign Ministry or the commander of a PLA Air Force base. But more senior leaders may not view the situation as a crisis. In addition, a military conflict in Southwest Asia will not have the same sense of urgency as a military conflict on the Korean Peninsula.

229

Moreover, cultural factors may lead Chinese leaders to think about crises differently than, say, American leaders. Chinese, for example, appear to view crises as not entirely negative phenomena. The Chinese term for crisis (*weiji*) is a combination of the words for danger (*weixian*) and opportunity (*jihui*).

With all this in mind, the editors decided to sidestep the crisis management minefield by adopting a rubric of "national security decisionmaking under stress."

Significant attention has focused on China's management of crises both inside the country and out. Whatever the rubric adopted in this volume, it is important to ask what the findings from this volume can tell us about Chinese handling of national security decisionmaking in times of great stress. Consequently, we will attempt to prepare a matrix demonstrating how the Chinese, and more specifically the PLA, respond to crises by type and importance.

So what constitutes a crisis? Using Jonathan Wilkenfeld's conception, we define a crisis as a situation that (1) presents a serious threat to the "basic values" or "core interests" of the actors involved; (2) involves a finite time or sense of urgency in the minds of the key actors; and (3) presents a key opportunity to advance or damage substantially the core interests of the key actors, including significant potential for military conflict.[1]

With so many variables, causes, degrees of severity, outcomes, and participants in the cases reviewed, it was a challenge to tie them all together. Moreover, because of the dearth of information available to the outside observer regarding China's decisionmaking processes and contemporary leadership dynamics for each case, no conception can claim total inclusiveness. Still, this treatment addresses the commonality observed in each case, with the presumption that a critical look at other crises will find a similar pattern. It is based not on any over-arching political science or foreign policy theory, nor does it really lend itself to traditional categories. It is purely based on direct observation, intuition (in the words of one participant), and in the belief that communist regimes tend to thrive in crises situations.

CRISIS MODE AS THE NORM?

One could make the argument that the Chinese Communist Party (CCP) *requires* crises for its continued existence. Communists came to power by convincing (or coercing) its constituents into believing that only the Communist Party could save them from the crisis in which they found themselves. The Communists built an "urban myth" around themselves that depicts them as the proverbial cavalry riding to the rescue just in time to save China from complete annihilation. This myth appears to be an important dimension in maintaining popular support for the regime. It is often said that the CCP has all but abandoned Marxism-Leninism and Maoism to justify its continued rule, and this has been replaced by "performance-based legitimacy." In other words, China's rulers rely very heavily on sustained economic growth and rising living standards for popular support. While this is so, this is only half of the story. The CCP leadership also leans very heavily on nationalism.[2] Crises provide invaluable opportunities for China's leaders to exploit deep and emotional groundswells of nationalism and conflate CCP rule with the power of and pride in the "new China." Hence the oft repeated mantra: "Without the Communist Party, there would be no new China" (*meiyou gongchandang, meiyou xin zhongguo*). Manufacturing and/or manipulating crises to whip up nationalist sentiment and national solidarity can be useful to China's leaders. There are three topics in particular that can really stir the passions of Chinese people: Taiwan, Japan, and the United States. Taiwan is the ultimate nationalist cause, and mere mention of efforts by "independence forces" on the island brings outpourings of popular indignation. Similarly, any perceived slight against China by unrepentant "Japanese militarists" or brazen "American hegemonists" can be counted on to produce expressions of outrage in Chinese on-line chat rooms and radio call-in shows. Thus, sometimes there are important reasons for the Party to create a crisis if one does not present itself.

A communist regime typically seized power in a protracted crisis, has operated for many years of its existence in a hyper crisis mode, and constructed a savior myth. As a result, a communist regime tends to be designed, organized, and its leadership trained to respond to crises. Therefore, communist leaders believe they need periodic crises to move policy forward and retain popular support.

WHY DOESN'T PRACTICE MAKE PERFECT?

So why, then, does the CCP often seem so inept at handling crises? The assertion that they thrive on crises presents a paradox. How can the regime thrive on crises and yet be incapable of managing them? This apparent paradox is key to appreciating the differences in how Beijing approaches crises. We suggest that it is useful to divide crises that China manages into three categories: fabricated, anticipated, and unanticipated. For fabricated crises (and partially for anticipated crises), China's leaders do have crisis management down to a science (but not an art). But they are so scientific in their approach that they cannot react in a flexible, swift, or artful way, explaining the time it takes to respond to real crises. For unanticipated crises, crisis management is far less impressive than it is for the former two categories (see Figure 1).

	Fabricated	Anticipated	Unanticipated
Type of Response	Coherent, decisive	Outward paralysis internal uncertainty	Graduated
Speed of Response	Swift, if not immediate	Initially very slow	Slow reaction
Unity of Action	Strong, cohesive well-coordinated	Imperfectly coordinated	Uncoordinated

Figure 1. Crisis Response Typology.

Moreover, handling crises in the post-Mao and post-Deng eras is not simple anymore. Management of crises in earlier eras was considerably more straightforward for the CCP leadership. Power tended to be much more centralized and located in a single individual—the paramount ruler. Decisions could be made by one man, and, once he gave the order, the crises were handled according to his instructions. In the era of Mao Zedong, and to a somewhat lesser extent in the era of Deng Xiaoping, the regime's performance in a crisis could be understood relatively easily, employing a unitary actor model. In the Jiang Zemin and the Hu Jintao eras, power is far more dispersed and bureaucratic politics far more important. One leading academic, Kenneth Lieberthal, has dubbed China's system in the post-Mao era as "fragmented authoritarianism."[3] In short, crisis management in the 21st century has become far more challenging than before for China's communist rulers.

CRISIS MANAGEMENT IN THE POST-DENG XIAOPING ERA

Because of the increasing challenge crisis management presents to Chinese leaders in the post-Deng era, there have been considerable efforts to develop coordination mechanisms to manage crises. There has been keen interest in establishing a Chinese version of a U.S.-style National Security Council. This was especially evident in the aftermaths of the Belgrade bombing episode of 1999 and the EP-3 Incident of 2001.[4] The Chinese trend to institutionalize crisis response was recently highlighted in a 2005 State Council report to the National People's Congress Standing Committee. In this case, a pre-packaged "counteremergency" response mechanism was proposed for the multitude of natural disasters that befall China annually.[5] Whether the Chinese government really believes it can perfect a response to any future crisis, the establishment of the nationwide contingency framework supports their need to *seem* as if they can. Some would argue that the government *must* be seen as in total control. Dr. Jiang Jinsong, in his seminal work on the National People's Congress, provides the historical basis for the Chinese concept of central power. According to Jiang, the lessons learned in the Spring and Autumn and Warring States periods included the concept that "a state under one supremacy will be orderly . . ."[6]

Perhaps the closest to an existing crisis management mechanism are the Leadership Small Groups (*lingdao xiaozu*) that exist to handle coordination between bureaucratic stovepipes.[7] These approximate what U.S. Government bureaucrats call "Interagency Working Groups."

GOALS IN A CRISIS

1. Survival of the Party.

First and foremost is the survival of the party. Absolutely no action will be taken that threatens the viability of their chosen political coattails upon which they believe the future of China rests. Puska breaks this priority into two parts—protecting the party's power and the party's reputation [Puska, 95]. Ensuring unchallenged rule of the CCP is accomplished through maintaining domestic stability and popular legitimacy. Given the importance of pressing forward with economic growth and reform, in the immediate aftermath of the Tiananmen crackdown it was critical to select a successor to Deng Xiaoping who could continue to push forward with economic development and had strong progressive/reformist credentials. Jiang Zemin met these requirements well: while he had demonstrated his loyalty to Deng, he was not viewed as a hardliner or conservative.

2. Enforce Party Unity.

The primacy of the first goal helps explain why an unanticipated crisis normally does not lead to competition between various entities for a solution, as party unity must be protected at all costs. The cashiering of Zhao Ziyang in May 1989 is a prime example of sacrificing a dissenter in order to preserve Party unity. This action indicated Party leaders saw the only way to weather the crisis was to enforce Party discipline and reassert Party unity. Wortzel's assertion that Zhao was using the Party's own organization department to attack the older leadership from within indicates that Zhao was indeed considered a direct threat to the party's core. [Wortzel, 61] Jiang Zemin was selected to replace Zhao in part because he adeptly and peacefully managed the protests in Shanghai, but also because he demonstrated his loyalty to Deng and the Party.

3. Protect China's International Credibility.

This goal is closely linked to the CCP's domestic credibility (see the first goal), but it remains important in its own right. In the case of Tiananmen, Jiang Zemin, the individual selected to succeed Deng Xiaoping, clearly had nothing directly to do with the use of lethal force in Beijing. Therefore he was untainted in the eyes of foreign governments who would have no reluctance to deal with him (in contrast to Premier Li Peng).

PHASES OF CRISIS MANAGEMENT

Study the Problem.

Considerable research and/or analysis is undertaken to create a strategy or game plan to determine how best to meet the goals listed above through management of the crisis. China is now in the post-Mao era where there is no longer a single all-powerful leader who decides everything. In a manufactured crisis, leaders may seek the advice of experts at various research institutes that have become important players in China's national security affairs over the past 3 decades.[8] The research step is a necessary one in a regime that rules by committee, and allows the decisionmakers to hear different opinions, recommendations, and the possible impact of each considered course of action on all concerned (affected) organizations. To an outside observer, however, this part of the process may resemble political "paralysis" [Wortzel, 56], when in all probability, it is better described as a blackout of information external to the concerned party organs. In this period, leaks should be viewed as either a breakdown in party discipline or a calculated trial balloon to test a potential response.

Devise a Strategy.

Develop a game plan and coordinate the implementation with different bureaucracies/*xi tongs* playing different roles. For example, on Taiwan the PLA played bad cop to the mainland's east coast provinces' good cop wooing of Taiwan investment to help finance their industrial modernization plans. These provinces have received at least U.S. $70 Billion in Taiwan investments, while as many as 75 new ballistic missiles have been deployed each year among their factories. According to official Chinese statistics, Taiwan investment in 2004 alone amounted to 9 U.S. billion dollars, while bilateral trade approached U.S. 78 billion dollars.[9] Finally, their research will focus on finding a solution to prevent the same crisis in the future. Understanding this step is critical in understanding the nature of crisis management in China. The real danger in a crisis is that the governing party will come under such blame by the public for the crisis itself that it loses its ability to govern. The idea is to survive the political crisis first, then handle the actual problem that created it. Shichor's representation of China's finding opportunity in the early Iraqi Wars without the presence of danger indicates they were comfortable operating within a crisis though they themselves were not working under crisis conditions. Only their observations of the unexpected U.S. military successes in the second and third Iraqi War created a crisis, as they pointed out serious deficiencies in the PLA's own modernization efforts.

Assign Blame Somewhere Else.

It is important to find a scapegoat for the crisis. It is also important to find a financier of the crisis recovery. Note that this logic is one step removed from the crisis facing the public. It is important to establish blame for the crisis somewhere other than on the shoulders of the Party. If the government can be spared as well, all the better, but the above Zhao example shows that even the government (and high-ranking Party leadership) is not immune if the Party is threatened. Often, this laying of blame is accompanied by an information operation (IO) to gain support for the decision. This IO is directed against the natural or intended antagonist in expected negotiations in an attempt to undercut the other party's negotiating positions vis-a-vis their own. Swaine referred to this substep as "shaping," meaning the Chinese will take steps to set favorable conditions for their own opening position.

The ultimate goal, to again use Swaine's words, is to create a *fait accompli*. For example, while Harlan Jencks' comment that the "PLA entering Tiananmen when ordered to do so is an indication of their increased professionalization in the late 1980s" is valid,[10] we would add that it also followed weeks of condition-setting (shaping) by the central leadership in which the deployed units were isolated from any external source of news and briefed that the capital was under siege. Wortzel gives an excellent example of troops being briefed of the conditions in the city over loudspeakers and using such Maoist class-warfare terms as "counter-revolutionaries" [Wortzel, 73]. This internal IO campaign continued after the events of June 3-4, with the awarding of each soldier involved a medal and watch inscribed "Liberator of the Capital," and the closing of Tiananmen Square until its subsequent cleaning and repair allowed history to be written on Beijing's terms. Chinese propaganda continues to insist that no one died in Tiananmen Square itself.

Nearly 12 years later, the U.S. Government's failure to counter China's IO campaign identifying the EP-3 as a spy plane illegally flying in Chinese airspace may have actually helped Beijing set the terms for how the world discusses that event. While the principal concern was to avert a hostage crisis [Godwin, 176], Washington's passiveness allowed China to establish the terms and conditions to such an extent that they continue to be used by even the American media today.[11] As Godwin also points out, Beijing's reluctance to follow international norms and laws for the quick return of the crew and aircraft was due to the central leadership's domestic and international need to be seen as tough on the United States. [Godwin, 183] Their delay in effect created a political crisis in Washington, which the Chinese used to their advantage.

Keep The Opponent Off Balance/Maintain the Initiative.

It is important to understand that none of this theory tries to explain the logic behind the decision made, though it probably is based in part on traditional teachings using modern methods. A study of the logical basis for Chinese decisionmaking would help in this regard. Richard Solomon notes that the absence of rationality in a Chinese negotiating position usually indicates factional political pressures.[12] Rational behavior is, of course, a subjective perception based on the observer's own experiences and logic. While factionalism may have much to do with this approach by the Chinese, equally possible is a calculated attempt to confuse their interlocutor or to force an attempt around the impasse from which the Chinese can better achieve their objectives. In so doing, the Chinese negotiator would force the western negotiator out of his planned approach, thereby creating a crisis of sorts that—while secondary to the main line of negotiations—must be dealt with first. Combined with a propensity of the Chinese to spend scheduled negotiating time establishing friendships with their interlocutors that would later be used to manipulate the friend's feelings, the Chinese easily can create the sense of a time crunch (crisis) toward the end of a negotiating period. Other pressures identified by Solomon as used by Chinese negotiators include leaks to the press, accusations of injury to China's prestige, word games, "killing the chicken to warn the monkey," and threats against favored Chinese officials.[13] All are designed to create a crisis in the mind of the interlocutor that, in turn, creates favorable conditions for the Chinese side.

Stack the Deck in Your Favor.

China's Communist leaders work very hard to create favorable conditions for achieving their goals. This is especially true for negotiations, and they undertake extensive preparations well before formal negotiations even begin. The idea that an external cause for a crisis must be found implies that negotiations will always be part of the Chinese crisis management calculus. Thus, the setting of conditions is critical to ensuring a favorable outcome. Historical examples are numerous to show that China uses all manner of techniques to put the other party in any future negotiations at a disadvantage, including cross-border attacks or other demonstrations of force.

Attacks in the open media are commonplace and serve to secure public opinion behind their position. Godwin's characterization of Chinese writings prior to the Belgrade Embassy bombing as increasingly anti-American [Godwin, 162-3] implies the creation of an opportunity waiting to happen. The Politburo decision to increase the public criticism of the United States and the North Atlantic Treaty Organization (NATO) can be seen as a worried Beijing setting the conditions in anticipation of a crisis they could use to slow a worrisome post-cold war U.S. strategy of intervention that marginalized China's position as a third pole. The accidental bombing of their embassy on May 7, 1999, gave them the opportunity they sought.

Signals are sent through an empathetic third party to the second negotiating party. This move not only gives China a credible witness to their warnings (vice direct contact), but it also removes the distraction of emotion between interlocutors/negotiators. China's record of success with this portion of their management plan is spotty. As Swaine implies, Asian countries are more likely to receive the signal as it was intended than Western countries. In a potential conflict scenario, signaling through a third party also ensures that it is China who makes the first international steps toward peaceful resolution.

Direct communications are also part of China's steps toward establishing favorable conditions. The primary objective prior to the start of formal negotiations is to get the other party to agree to a set of preconditions (or "principles") which will be used later to measure whether negotiations by the other party are sincere. Refusal by the other party to accept China's preconditions will invoke a series of increasingly direct and public attacks on that party to increase domestic and international pressure on it to enter negotiations. President George Bush points this out with regards to Beijing's hammering away at Chen Shuibian"s refusal to accept the "One China" principle. However, whether China may be willing to set that precondition aside at some time in the future is debatable [Bush, 150]. Godwin also identified this phenomenon in his study of the EP-3 crisis, noting that China's negotiating tactics became "stiffer" following unconciliatory statements by President Bush and Secretary of State Colin Powell. Japan is facing this tactic at the moment regarding the chemical weapons left behind in 1945, with the Chinese goal of engaging the Japanese in negotiations to accept responsibility (giving them greater leverage over Japan in other negotiations) and to pay for the cleanup and compensations.

THE PLA AND THE CRISIS TYPOLOGY

Not all crises are created equal, and therefore the speed and manner to which they are reacted is dependent on their type. For simplicity, this theory categorizes crises into three types: fabricated, anticipated, and unanticipated. Each category has visibly distinct reactions to an observer looking for the right indicators, providing a relatively high potential of identifying the type of crisis (from the Chinese point of view) before entering into negotiations or deciding on some other course of action. In general terms, when faced with a planned or anticipated crisis, the government will act with speed and decisiveness. The difference between planned and anticipated is in the degree of unity with which the government acts (see Figure 1).

The anticipated category allows for differences of preparedness between the various ministries or other sub-elements of the government. In the case of an unplanned or real crisis, however, reactions are very different. Chinese reactions to a real crisis are very slow. In the words of one conference participant, they "circle the wagons," [Puska, 95] instituting various study groups to research the issue before determining what position—and what actions—to take.

The PLA and Fabricated Crises.

The perfect conditions for the Chinese are situations in which they can exert complete control and can create and manipulate a crisis in the minds of their target audience. Such was allegedly the case for the Taiwan Strait crisis of 1996. According to participants of Harvard's Senior Executive Course in March 1998 who claim to have been personally involved in managing the crisis, the whole purpose of the missile firings was to prompt a two-carrier response by the United States. Anything less, they commented, would have been a disappointing indication that Washington did not get the intended message.[14] This is perhaps *ex-post facto* wisdom and bravado, as most Chinese civilian and military analysts one author spoke with in Beijing and Shanghai in 1998 about the Strait crisis said that Chinese leaders were caught completely off-guard by the U.S. response—they were quite simply shocked![15] If the latter is the actual case, the delegates at Harvard must have wanted to appear in control, unless they were victims of their own IO spin. Regardless, the Taiwan Strait crisis of 1996 was *planned* by the PRC to create a political crisis for the Taiwan national elections. There is little evidence to suggest they strayed from their game plan during this period, even after the heavy U.S. response.[16] This crisis is therefore marked as fabricated in Figure 2.

	Fabricated	Anticipated	Unanticipated
Taiwan (95-96)	X		
SARS		X (PLA)	X (MOH/Beijing)
Tiananmen Sq	X PLA	X (CCP Elders)	
EP-3			X
Belgrade Bombing			X
Persian Gulf crises		X	

Figure 2. Crisis Response Typology.

The PLA's role in crisis negotiations depends in large part on the location of the other party. During many of the country's border negotiations with its neighbors, the PLA was often used to send a show of resolve to the other party. At times, this was direct, as in the case of the fight with the Soviet Union over several islands in the Ussuri River. At other times, it was indirect, such as using the border skirmishes against India in 1962 to pressure Burma to acquiesce to China's demands.[17] The PLA can also be repositioned as a warning of China's seriousness. The buildup of surface-to-surface missiles across from Taiwan is discussed openly by the Chinese in this manner. Rumors of PLA repositioning of forces along the North Korean border may also have been meant to signal Pyongyang not to pull out of the six-party talks. Given recent rumors of instability in Pyongyang, these PLA redeployment rumors may also have been targeted as a warning to any domestic Korean threat to Kim Jong il's leadership.

The PLA and Unanticipated Crises.

In the case of the severe acute respiratory syndrome (SARS) crisis, however, the PLA played a much different role. As demonstrated through Puska's in-depth analysis of this completely unanticipated crisis, the PLA was the first Chinese organ to break the official silence and take action to treat the outbreak in a cohesive manner. Perhaps the best example provided was of the relative speed and decisiveness shown by the PLA Medical Department's response to the SARS outbreak in Beijing, long before the Ministry of Health and the Beijing City Government reversed their policies of denial.

This indicates, perhaps, a different set of goals for the PLA than for the Party, and supports the idea that China is not a unitary actor in all cases. We would argue that, while the SARS epidemic was truly a crisis for most of the Chinese leadership, the PLA had decided post-Tiananmen to take advantage of every opportunity to redeem itself in the eyes of the Chinese people.

The PLA and Anticipated Crises.

The flooding of the Yangzi River in 1998 offered the best example of a clear campaign by the PLA not just to help contain the damage being done by nature, but to do so in full view of the people in an attempt to regain their support. In other words, the PLA anticipated the crisis and seized this as an opportunity to publicly demonstrate its connection to the people. SARS allowed them another opportunity, and again they took it, this time at the expense of other government organs—the Ministry of Health and the Beijing City Government. The PLA was able to act faster than these other organizations by redefining the nature of the crisis to highlight the opportunity over the danger. Only later were other entities ready to play a part, but by that time the PLA had taken the credit for saving the day, creating in this manner a separate crisis in party unity that was only solved a year later when the PLA's hero during the SARS crisis, Dr. Jiang Yanyong, was arrested for publicly criticizing the Tiananmen decisions and calling for an open apology by the Party.[18]

By looking at Chinese crisis management in this manner, actions previously seen as illogical or beyond understanding seem to make sense. In the spring of 1989, the leadership was prepared to act only when international events hosted by Beijing (first the Asian Development Bank meeting and then the first Sino-Soviet summit in 30 years) had all concluded. There were, therefore, no distractions to prevent the leadership from focusing on the issue of getting the students back into the classrooms before a second Cultural Revolution occurred. Once the leadership was ready to act, the crisis allowing them to do so was orchestrated and publicly portrayed. Observers have often opined that many of the actions by the PLA and People's Armed Police (PAP) in late May and early June seem to have been an attempt to pick a fight. The attempted infiltration of troops in civilian clothes and the later reports that many of the vehicle fires were set by their own drivers both point to increasingly direct attempts to place the troops in harm's way, waiting for an armed or violent response by the students.[19] When that did not happen, was it fabricated?

One participant"s recollection of PLA Commanders' feelings following the actions of June 3-4 and another participant"s discussion of a possible "Duality of Command" make one wonder if "dual orders" were issued intentionally on that fateful weekend to foment the appearance of chaos and hence provide justification for a crackdown. To be specific, was there deliberate differentiation between the instructions given to those already deployed in inner-city areas and those units outside that were given the orders to "retake" the city to restore order. Was all of this part of a larger plan to set the favorable conditions necessary to get units of the People's Army to take action against the people? Wortzel's observations of inner-city PAP bulletin boards support this possibility, and it certainly fits the model. One author's research indicates that Academy of Military Sciences and National Defense University students were allowed to join the mass of civilians that delayed for 3 days one unidentified armor unit from crossing the western canal bridges near their campuses.[20]

CONCLUSION

If Swaine's assertion is true that China believes the United States will choose to avoid a crisis of force with China, then it suggests the possibility that China will seek to create crises in dealing with the United States. [Swaine, 18] For this reason, if no other, the United States must identify the methods by which China creates crises, and then acts to "resolve" them. When analyzing a crisis, both the dangers and the opportunities must be identified for all players and at all levels. Linear decisionmaking cannot be assumed, nor can China be considered a unitary actor. This implies the need to identify seams that can be exploited (as well as working to close the seams between the various players on the U.S. side). China's center of gravity is clearly identified as the Party. Above all things, the continued prominence of the party will be protected. Attacking this center of gravity in a crisis, therefore, is likely to create a unitary response and is probably not in the U.S. interest.

Determining whether the crisis being presented is fabricated, anticipated, or unanticipated is crucial. The value of the crisis typology identified in this chapter (see Figure 1) is that it allows one to clarify key crisis characteristics and permit swift identification of crisis type. Once this is determined, the other party can then better decide its own courses of action. If it is unanticipated, then in the minds of Chinese leaders, the crisis threatens the continued existence of the Party. If fabricated, the "crisis" may be designed to push forward an agenda or to deflect attention away from a real crisis.

A fabricated crisis is indicated by swift, decisive action (Figure 1). An unanticipated crisis is indicated by a "circling of the wagons," loud and repeated messages of an initial party line, while tangential actions are taken to gain control or divert attention. Only later will actions be taken against the original crisis. This is due to the fact that a "real" crisis implies surprise, with no preparation or plan on how to react. This situation requires studying, with the inherent tendency of communist study processes to be slow because of the need to allow every faction a say in the proceedings.

As Swaine pointed out, do not expect a tit-for-tat approach to a crisis [Swaine, 19]. The Chinese feel they will lose control if they become reactive in nature. The decisionmaking system does not allow for real or near-real time decisions. The idea of creating the perfect condition before entering the fray supports the principle that China does not "respond" to crises, if given the choice.

Is there an institutional explanation for why some crises receive quicker responses than others? Perhaps the PLA is just better organized to anticipate and plan for crises. As Godwin points out, the PLA was not directly involved in the negotiations that followed the Belgrade bombing and aircraft collision. [Godwin, 186] There can be no doubt, however, that the PLA was closely involved in the latter, perhaps even to the extent that internal Chinese negotiations were taking place in parallel to the international set. Perhaps it was the capability of the PLA, as an institution and as represented by the Central Military Commission (CMC), to conclude their position and draw their negotiating lines before the Ministry of Foreign Affairs or the Politburo Standing Committee could decide the national position.[21] If so, this capability for the PLA to get ahead of policy in the event of a crisis bodes negatively in any future increase of tensions over Taiwan, the Diaoyutai/Senkaku Islands, or on the Korean Peninsula. If the trend identified by Robert Suettinger continues, this could equate to military action *in extremis* of a policy decision, placing everyone in a crisis mode. [Bush 148, *n*12] The recently legislated Anti-Secession Law may actually make this more of a danger in the future.

Whether the topic is dubbed "decisionmaking under stress" or "crisis management," this volume represents only a first cut. More research is needed desperately on this subject. Analyses of other case studies that examine the roles of key actors in the Chinese national security establishment, including the PLA, are essential. This is especially true where Taiwan and the Korean Peninsula are concerned. A case study of the drafting and passage of the Anti-Secession Law should be made a priority. The bill, which became law in March 2005, appears to be an instance of Beijing attempting to fabricate a crisis. What are the goals of the regime here? What was the role of the PLA in generating this legislation? Regarding the Six-Party talks on North Korea, what was the genesis and evolution of this initiative, which brought the United States, North Korea, and other concerned parties to the same table? The story behind the 2003 launching of these talks would provide fascinating insight into how Beijing handles complex and pressing challenges on its periphery.

ENDNOTES - CHAPTER 8

1. Jonathan Wilkenfeld, "Concepts and Methods in the Study of International Crises," Unpublished paper, College Park: University of Maryland, February 2004. See also Michael Brecher and Jonathan Wilkenfeld, *A Study of Crisis*, Ann Arbor, MI: University of Michigan Press, 2000.

2. For a recent treatment of Chinese nationalism, see Peter Gries, *China's New Nationalism: Pride, Politics, and Diplomacy*, Berkeley and Los Angeles: University of California Press, 2004.

3. Kenneth Lieberthal, *Governing China: From Revolution Through Reform*, New York: W. W. Norton, 1995, pp. 169-170. See also Lieberthal's chapter in David M. Lampton and Kenneth Lieberthal, eds., *Bureaucracy, Politics, and Decisionmaking in Post-Mao China*, Berkeley and Los Angeles: University of California Press, 1992.

4. See, for example, Charles Hutzler, "China's Heir Apparent May Encounter Turf Battles," *Wall Street Journal*, April 30, 2002, p. A15.

5. Ling Hu, "Crisis Countermeasures Drawn Up," *China Daily*, Weekend Edition, February 26-27, 2005, p. 1.

6. Jiang Jinsong, *The National People's Congress of China*, Foreign Languages Press, Beijing, China, January 2003, p. 6. Jiang was quoting the writings of Xun Kuang.

7. On this subject, see Taeho Kim, "Leading Small Groups: Managing All Under Heaven," in David M. Finkelstein and Maryanne Kivlehan, eds., *China's Leadership in the 21st Century: The Rise of the First Generation*, Armonk: M. E. Sharpe, Inc, 2003, pp. 121-139.

8. For more on the role and importance of Chinese think tanks, see Evan Medeiros, "Agents of Influence: Assessing the Role of Chinese Foreign Policy Research Organizations after the 16th Party Congress," in Andrew Scobell and Larry Wortzel, eds., *Civil-Military Change in China: Elites, Institutes, and Ideas After the 16th Party Congress*, Carlisle Barracks, PA: U.S. Army War College, Strategic Studies Institute, 2004, pp. 279-308; and *China Quarterly*, Issue No. 171, September 2002, which contains several articles on Chinese think tanks.

9. The 70 billion figure comes from "China-Taiwan Ties Strengthen Despite Political Unease," *OxResearch*, January 31, 2005, p.1. The 2004 figures were reported by a China Radio International broadcast on March 31, 2005.

10. Harlan W. Jencks, "Civil-Military Relations in China: Tiananmen and After," *Problems of Communism*, Vol. XL, No. 3, May-June 1991, pp. 14-29.

11. Hear, for example, on any news broadcast regarding this event, how "the American spy plane collided with the Chinese fighter," instead of how the Chinese fighter collided with the American reconnaissance plane.

12. Richard H. Solomon, *Chinese Political Negotiating Behavior: A Briefing Analysis*, Santa Monica, CA: RAND, December 2005, p. 4.

13. Solomon, pp. 10-14.

14. Susan A. Lawrence, *Impressions of the Second Kennedy School Executive Program for Senior Chinese Military Officers*, March 7-21, 1998, unpublished, p. 31.

15. See Andrew Scobell, *China's Use of Military Force: Beyond the Great Wall and the Long March*, New York: Cambridge University Press, 2003, p. 182. See also John W. Garver, *Face Off: China, the United States, and Taiwan's Democratization*, Seattle: University of Washington Press, 1997, chapters 9 and 10.

16. However, it does appear that the PLA tested fewer missiles in March 1996 for some reason. Scobell, *China's Use of Military Force*, p. 183.

17. Frank Miller, unpublished notes from the joint Atlantic Council—Asia Pacific Center Conference on Asian Approaches to International Negotiations: Borders and Territories, Honolulu, HI, September 8-10, 1998.

18. Dr. Jiang's subsequent arrest for criticizing the Party may help explain why he broke ranks (thereby creating an exception for this model) with the Party during the SARS crisis. It stands as a good example, though, of how intra-Party debates are handled when they become public.

19. Scobell, *China's Use of Military Force*, pp. 162-163.

20. Miller, Discussion with individuals who were National Defense University students in 1989, Beijing, China, November 2003.

21. For details on the extraordinary powers of the PLA, through the Central Military Commission, to propose laws and present its views to the central leadership faster than any other government entity, see Jiang, *The National People's Congress of China*, pp. 137-140. For the CMC's role in decisionmaking, see Tai Ming Cheung, "The Influence of the Gun: China's Central Military Commission and Its Relationship with the Military, Party, and State Decisionmaking Systems," in David M. Lampton, ed., *The Making of Chinese Foreign and Security Policy in the Era of Reform*, Stanford, CA: Stanford University Press, 2001, pp. 191-122.

ABOUT THE CONTRIBUTORS

RICHARD BUSH is a Senior Fellow at the Brookings Institution and Director of its Center for Northeast Asian Policy Studies. The Center serves as a locus for research, analysis, and debate to enhance policy development on the pressing political, economic, and security issues facing Northeast Asia and U.S. interests in the region. Dr. Bush came to Brookings in July 2002, after serving almost 5 years as the Chairman and Managing Director of the American Institute in Taiwan (AIT), the mechanism through which the U.S. Government conducts substantive relations with Taiwan in the absence of diplomatic relations. He began his professional career in 1977 with the China Council of The Asia Society. In July 1983, he became a staff consultant on the House Foreign Affairs Committee's Subcommittee on Asian and Pacific Affairs. In January 1993, Dr. Bush moved up to the full committee, where he worked on Asia issues and served as liaison with Democratic members. In July 1995, he became National Intelligence Officer for East Asia and a member of the National Intelligence Council (NIC), which coordinates the analytic work of the intelligence committee. He left the NIC in September 1997 to become head of AIT. Dr. Bush is the author of a number of articles on U.S. relations with China and Taiwan, including *At Cross Purposes*, a book of essays on the history of America's relations with Taiwan (Armonk, NY: M. E. Sharpe, March 2004), and his study on cross-Strait relations, *Untying the Knot: Making Peace in the Taiwan Strait* (Washington, DC: Brookings, forthcoming 2005). Dr. Bush received his undergraduate education at Lawrence University in Appleton, Wisconsin. He did his graduate work in political science at Columbia University, New York, receiving an M.A. in 1973 and his Ph.D. in 1978.

PAUL H. B. GODWIN is a Senior Fellow at the Foreign Policy Research Institute, Philadelphia, Pennsylvania. He retired as professor of international affairs, the National War College, Washington, DC, in the summer of 1998. In the fall of 1987, he was a visiting professor at the Chinese National Defense University. His teaching and research specialties focus on Chinese defense and security policies. Dr.

Godwin's most recent publications are "Change and Continuity in Chinese Military Doctrine: 1949-1999," in Mark A. Ryan, David M. Finkelstein, and Michael A. McDevitt, eds., *Chinese Warfighting: the PLA Experience Since 1949* (Armonk, New York: M.E. Sharpe, 2003); "China's Defense Establishment: The Hard Lessons of Incomplete Modernization," in Laurie Burkitt, Andrew Scobell, and Larry Wortzel,eds., *The Lessons of History: The Chinese People's Liberation Army at 75* (Carlisle, PA: U.S. Army War College, Strategic Studies Institute, 2003); *Preserving the PLA's Soul: Civil Military Relations and the New Generation of Chinese Leadership*, CAPS Papers No. 33 (Taipei, Taiwan, Republic of China: Chinese Council of Advanced Policy Studies, October 2003); and "China as Regional Hegemon?" in Jim Rolfe (ed.) *The Asia-Pacific Region in Transition* (Honolulu, Hawaii: Asia-Pacific Center for Security Studies, 2004). Dr. Godwin graduated from Dartmouth College with a degree in International Relations, and received his Ph.D. in Political Science from the University of Minnesota.

JAMES R. LILLEY is Senior Fellow in Asian studies at the American Enterprise Institute. Mr. Lilley was the U.S. Ambassador to the People's Republic of China from 1989 to 1991 and to the Republic of Korea from 1986 to 1989. He served as Assistant Secretary of Defense for International Security Affairs from 1991 to 1993. Mr. Lilley wrote the forewords for the AEI publications *Chinese Military Modernization, Over the Line,* and *China's Military Faces the Future.* He is also the coeditor of *Beyond MFN: Trade Beyond MFN: Trade with China* and *American Interests and Crisis in the Taiwan Strait.* He is the author most recently of: *China Hands: Nine Decades of Adventure, Espionage, and Diplomacy in Asia* (with Jeffrey Lilley), New York: Public Affairs, 2004.

FRANK MILLER is currently the Army Attaché to the U.S. Embassy in Beijing. As a Special Forces Officer assigned to First Special Forces Group and as a China Foreign Area Officer, he has served throughout East Asia, including tours in Okinawa, Hong Kong, and Hawaii as the J5 Country Manager for China, Taiwan, Hong Kong, and Mongolia. His previous tour was as the Defense and Army Attaché to the Socialist Republic of Vietnam in Hanoi. Colonel Miller is a graduate of the U.S. Military Academy, Naval Postgraduate School, and the U.S. Army War College.

SUSAN M. PUSKA retired from the U.S. Army on February 1, 2005. As a military logistician, she served in various command and staff positions in Germany, Korea, Fort Hood, Texas, and Aberdeen Proving Ground, Maryland. Her last position prior to retirement was Political-Military Advisor, Bureau of Asian and Pacific Affairs, Department of State. Prior to this she served as the Army Attaché, U.S. Embassy, Beijing, during 2001-03. During 1999-2000, Colonel Puska was the Director of Asian Studies, Department of National Security and Strategy, U.S. Army War College. Between 1996 and 1999, she served as the China Political-Military Desk Officer, Office of the Deputy Under Secretary of the Army, International Affairs. During 1992 to 1994, she was an Assistant Army Attaché at the U.S. Embassy, Beijing. Her publications include "SARS 2002-2003: A Case Study in Crisis Management," in *Chinese Crisis Management* (Carlisle, PA: Strategic Studies Institute, forthcoming 2005); "The People's Liberation Army General Logistics Department: Toward Joint Logistics Support," in *PLA As Organization* (Santa Monica, California: RAND, 2002); editor of *People's Liberation Army After Next* (Carlisle, Pennsylvania: Strategic Studies Institute, 2000); and *New Century, Old Thinking: The Dangers of the Perceptual Gap in U.S.-China Relations*, (Carlisle, Pennsylvania: Strategic Studies Institute, 1999). She studied Mandarin Chinese at the Defense Language Institute, Monterey, California, and The John Hopkins-Nanjing University Center for Chinese and American Studies, Nanjing, PRC. Colonel Puska has a BA from Michigan State University, an M.A. from the University of Michigan, and is a 1997 graduate of the U.S. Army War College, Carlisle, Pennsylvania.

ANDREW SCOBELL is Associate Research Professor at the U.S. Army War College Strategic Studies Institute and Adjunct Professor of Political Science at Dickinson College, both located in Carlisle, Pennsylvania. Dr. Scobell's research focuses on political and military affairs in the Asia-Pacific Region. He is the author of *China's Use of Military Force: Beyond the Great Wall and the Long March* (Cambridge

University Press, 2003). Dr. Scobell earned a Ph.D in political science from Columbia University.

YITZHAK SHICHOR is Professor of East Asian Studies and Political Science and Acting Chairman of the Department of East-Asian Studies at the University of Haifa. He is also a Senior Fellow at the Harry S Truman Research Institute for the Advancement of Peace at the Hebrew University where he served as Executive Director. Before his retirement from the Hebrew University in 2002, he had held the Michael William Lipson Chair in Chinese Studies and served as Dean of Students and Chairman of the Department of East Asian Studies and the Department of Political Science. Dr. Shichor is the author of *The Middle East in China's Foreign Policy, 1949-1977* (Cambridge: Cambridge University Press, 1979) and *East Wind over Arabia: Origins and Implications of the Sino-Saudi Missile Deal* (Berkeley: University of California Press, 1989). Dr. Shichor's articles dealing with Chinese military affairs, China's Middle East policy, China's international energy policy, and Muslim unrest in Xinjiang have appeared in *Asian Survey*, *The China Quarterly*, *Problems of Communism*, *China Report*, *Survival*, and *Pacifica Review*.

MICHAEL D. SWAINE came to the Carnegie Endowment after 12 years at the RAND Corporation. He specializes in Chinese security and foreign policy, U.S.–China relations, and East Asian international relations. At RAND, he was a senior political scientist in international studies and also research director of the RAND Center for Asia-Pacific Policy. Dr. Swaine was appointed as the first recipient of the RAND Center for Asia-Pacific Policy Chair in Northeast Asian Security, in recognition of the exceptional contributions he has made in his field. Prior to joining RAND in 1989, Dr. Swaine was a consultant with a private sector firm; a postdoctoral fellow at the Center for Chinese Studies, University of California at Berkeley; and a research associate at Harvard University. He attended the Taipei and Tokyo Inter-University Centers for Language Study, administered by Stanford University, for training in Mandarin Chinese and Japanese. One of the most prominent U.S. analysts in Chinese security studies, he is the author of more than 10 monographs on security policy in the region. Some of his publications are "US Security Policy under Clinton and Bush: China and Korea" in Chae Jin Lee, ed., *U.S. Security Policy*

Under Clinton and Bush: Continuity and Change, Claremont: The Keck Center for International Affairs, 2004); "Taiwan's Defense Reforms and Military Modernization Program: Objectives, Achievements, and Obstacles" in Nancy Bernkoppf Tucker, ed., *Dangerous Strait* (New York, Columbia University Press, 2005); "China: Exploiting a Strategic Opening," in Ashley J. Tellis and Michael Wills, eds., *Strategic Asia 2004-2005* (Seattle: National Bureau of Asian Research, 2004); "Deterring Conflict in the Taiwan Strait," *Carnegie Paper* No. 46 (July 2004); "The Taiwan Relations Act: The Next 25 Years," Testimony to the U.S. House of Representatives Committee on International Relations, April 2004; "Trouble in Taiwan," *Foreign Affairs*, March/April 2004. Dr. Swain received a B.A. from George Washington University and an A.M. and Ph.D. from Harvard University.

LARRY M. WORTZEL is a commissioner on the U.S.-China Economic and Security Review Commission, a Congressionally-appointed body. He is also a fellow with The Heritage Foundation, where he previously was Vice President for Foreign Policy and Director of the Asian Studies Center. Dr. Wortzel was a career Army officer, retiring as a colonel after 32 years of military service in 1999. He spent 7 years as an infantryman. He has long experience as a counterintelligence officer and foreign intelligence collector for the Defense Department. In 1971-72, while assigned in Southeast Asia, he collected communications intelligence on Chinese forces in Laos and Vietnam. He was an intelligence analyst on China at the U.S. Pacific Command from 1978-82. Dr. Wortzel served as a military attaché in China from May 1988 to June 1990 and June 1995 to December 1997. He also served as a strategist on the Department of the Army staff. His last military position was Director of the Strategic Studies Institute, U.S. Army War College. Dr. Wortzel is the author of two books on military history and political issues in China and has edited and contributed chapters to five other books on the Chinese armed forces. He has been quoted widely on these subjects in newspapers and on wire services. He studied at the Chinese Language and Research Centre of the National University of Singapore in 1982-83. A graduate of the Armed Forces Staff College and the U.S. Army War College, Dr. Wortzel earned his B.A. from Columbus College, Georgia, and his M.A. and Ph.D. from the University of Hawaii.